The High Energy Universe
Ultra-High Energy Events in Astrophysics and Cosmology

In the last two decades, cosmology, particle physics, high energy astrophysics, and gravitational physics have become increasingly interwoven. The intense activity taking place at the intersection of these disciplines is constantly progressing, with the advent of major cosmic ray, neutrino, gamma-ray, and gravitational wave observatories for studying cosmic sources, along with the construction of particle physics experiments using beams and signals of cosmic origin.

This book provides an up-to-date overview of the recent advances and potential future developments in this area, discussing both the main theoretical ideas and experimental results. It conveys the challenges, but also the excitement associated with this field. Written in a concise yet accessible style, explaining technical details with examples drawn from everyday life, it will be suitable for undergraduate and graduate students, as well as for other readers interested in the subject. Color versions of a selection of the figures are available at www.cambridge.org/9780521517003.

PÉTER MÉSZÁROS is Eberly Chair of Astronomy & Astrophysics and Professor of Physics at the Pennsylvania State University, where he is also Director of the Center of Particle Astrophysics. His main research interests are high energy astrophysics and cosmology. He has been a co-recipient of the Rossi Prize of the American Astronomical Society and the First Prize of the Gravity Research Foundation. He is a member of the American Academy of Arts and Sciences and the Hungarian Academy of Sciences.

The High Energy Universe

Ultra-High Energy Events in Astrophysics and Cosmology

PÉTER MÉSZÁROS

Pennsylvania State University

CAMBRIDGE
UNIVERSITY PRESS

CAMBRIDGE UNIVERSITY PRESS
Cambridge, New York, Melbourne, Madrid, Cape Town, Singapore,
São Paulo, Delhi, Dubai, Tokyo, Mexico City

Cambridge University Press
The Edinburgh Building, Cambridge CB2 8RU, UK

Published in the United States of America by Cambridge University Press, New York

www.cambridge.org
Information on this title: www.cambridge.org/9780521517003

First published 2010

Printed in the United Kingdom at the University Press, Cambridge

A catalog record for this publication is available from the British Library

ISBN 978-0-521-51700-3 Hardback

Additional resources for this publication at www.cambridge.org/9780521517003

Deborahnak, Andornak

Contents

Preface

This book provides an overview of topics in high energy, particle and gravitational astrophysics, aimed mainly at interested undergraduates and other readers with only a modest science background. Mathematics and equations have been kept to a minimum, emphasizing instead the main concepts by means of everyday examples where possible. I have tried to cover and discuss in some detail all the major areas in these topics where significant advances are being made or are expected in the near future, with discussions of the main theoretical ideas and descriptions of the principal experimental techniques and their results.

Cosmology, particle physics, high energy astrophysics and gravitational physics have, in the last two decades, become increasingly closely meshed, and it has become clear that thinking and experimenting within the isolated confines of each of these disciplines is no longer possible. The multi-channel approach to investigating nature has long been practiced in high energy accelerators involving the strong, the weak and the electromagnetic interactions, whereas astrophysics has long been possible only using electromagnetic signals. This situation, however, is rapidly changing, with the advent of major cosmic-ray, neutrino and gravitational wave observatories for studying cosmic sources, and the building of particle physics experiments using beams and signals of cosmic origin. At the same time, theoretical physics has increasingly concentrated efforts in attempts to unify gravity with the other three forces into an ultimate theory involving all four. The intense activity in these fields is beginning to open new vistas onto the Universe and our understanding of Nature's working on the very small and very large scales. In this book I have sought to convey not only the facts but also the challenges and the excitement in this quest.

I have been fortunate in my collaborators working in these fields and, at my own university, in having colleagues active in the various areas discussed here. Among the latter, I am grateful to Irina Mocioiou, Yuexing Li, Niel Brandt, Michael Eracleous, Derek Fox, Abe Falcone, L. Sam Finn, Paul Sommers, Douglas Cowen and Stephane Coutu for providing me feedback and advice on individual chapters. I am also grateful to my wife Deborah for suggestions on improving the readability of the manuscript. Any remaining errors are my own.

Understanding our cosmic environment and its immense displays of power is somewhat akin to experiencing a major storm at sea. One feels awe at its vastness and violence, and also the desire to understand, as far as possible, how it works and what causes it. I hope that this book will help its readers participate in this experience.

1

Introduction

1.1 The dark and the light

The Universe, as we gaze at it at night, is a vast, predominantly dark and for the most part unknown expanse, interspersed with myriads of pin-pricks of light. When we consider that these light spots are at enormously large distances, we realize that they must be incredibly bright in order to be visible at all from so far away. Occasionally, some of these specks of light get much brighter, and some of them which were not even seen with the naked eye before become in a few days the brightest spot in the entire night sky, their brightness having increased a billion-fold or more against the immutable-looking dark background. Thus, we have come to realize that the Universe is characterized by what Renaissance artists called *chiaroscuro*, referring to the contrast between light and dark, which is both stark and subtle at the same time. In the case of the Universe, the contrasts can be enormous and surprisingly violent, as well as of a subtlety which beggars the imagination. In this book we will focus on these contrasts between the vast, unknown properties of the dark Universe and its most violent outpourings of energy, light and particles.

According to current observations and our best theoretical understanding, the Universe is made up of different forms of mass, or rather of mass-energies, since as we know from special relativity, to every mass there corresponds an energy $E = mc^2$ and vice versa, where E is energy, m is mass and c is the speed of light. About 74% of the Universe's total energy content is in the form of *dark energy*, a very strange component whose true nature we are completely igno-rant of. All we know about it at present is what it appears to do to us and to the rest of the massive objects in the Universe: it affects the rate of the expan-sion of the Universe. The next most prominent mass-energy component in the

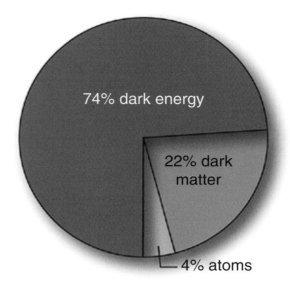

Figure 1.1   Relative amount of different forms of mass-energy densities in the Universe.
Source: SNAP project website.

Universe, amounting to about 22% of the total, is in the form of *dark matter* (another "dark" constituent!), of whose nature we are only slightly less ignorant than we are about dark energy. Despite 30 years of pondering it, all we know for sure about dark matter is how it affects the gravitational attraction felt by the "normal" matter of galaxies, we know roughly how it is distributed in space, and we can rule out some classes of objects as being responsible for it. The remaining fraction of the mass-energy of the Universe amounts to 4%, which is in the form of "normal" everyday baryons, or atoms (Fig. 1.1), although only about one in 10 of these (∼ 0.5%) emit light or detectable radiation, a very modest-looking contribution indeed. Physicists have taken to describing these two types of components as the dark and the light sectors of the Universe.

1.2   Where the fires burn

In the deep dark night of the Universe, the tiny bright specks of light shine as reassuring outposts, or so it would seem. These small corners of the Universe where we feel warm and at home form that portion which we *can* probe with our various instruments, telescopes, satellites, accelerators and laboratory experiments. In fact, this portion of the Universe makes up for its relatively small size with its sheer brilliance, and upon closer inspection, with its concentrated violence.

The most obvious denizens of the light sector, just from their sheer numbers, are the so-called main sequence stars, of which the Sun is a very ordinary example. The Sun's luminosity, that is its energy output per unit time, is $L \simeq 4 \times 10^{33} \, \mathrm{erg\,s^{-1}} \equiv 4 \times 10^{27}$ watts, which can also be expressed as 5×10^{23} horsepower.[1] Most of this energy, in the case of the Sun, is in the form of "optical" light, to which our eyes are sensitive, with smaller fractions in the infrared and in the ultraviolet parts of the electromagnetic spectrum. There are other stars which emit most of their electromagnetic radiation outside the optical range, either at shorter or longer wavelengths. Like the Sun, all stars shine because of nuclear reactions going on in their core, which results in their emitting copious amounts of neutrinos, a type of elementary particle, the stellar neutrino luminosity being in general comparable to the electromagnetic luminosity.

Despite their huge power, stars are just the lumpen proletariat of the Universe, humble light-bugs compared to some of the rare, lavish energy plutocrats which arise occasionally here and there. When they occur, the sky is pierced by extremely concentrated outbursts of high energy radiation pouring out from them, which make the normal stars pale by comparison, outshining them by a factor of a billion or more over periods of weeks. These outlandish events are called supernovae, and besides their optical and other forms of electromagnetic radiation, we have managed to measure on at least one occasion their neutrino luminosity as well. They are also thought to be powerful sources of other forms of cosmic rays, and to a lesser degree of gravitational waves, which however have not so far been detected. Some of these supernovae occur as a consequence of the collapse of the inner core of massive stars, while others are due to smaller stars slowly gaining mass until a nuclear deflagration occurs. In many cases, the collapse leaves behind an extremely compact remnant called a neutron star, composed of matter whose density is extremely high, comparable to that of atomic nuclei.

The most extreme stellar outbursts, however, appear to occur as a result of the core collapse of the most massive stars leading to the formation of a black

---

[1] We use the common scientific notation where a quantity written as, say, 6×10^X is equivalent, in the usual decimal notation, to 6 followed by X zeros before the decimal point, for instance, $6 \times 10^3 \equiv 6000$, or in general, the first number followed by X figures, with zeros added after the significant figures to make up X figures after the first one, for instance, $1.56 \times 10^4 = 15\,600$.

hole, or as a result of the merger of two compact stars leading to a black hole. The black hole formation may perhaps proceed through an intermediate stage as a neutron star with an extremely high magnetic field. These cataclysmic events are called "gamma-ray bursts", or GRBs. They flare up very fast, and for short periods of time (seconds or minutes), their brightness can exceed the total luminosity of the rest of the observable Universe.

Slower flares of even higher total energy occur in some galaxies, made up of billions or trillions of stars. These are related to massive black holes which lurk at the center of most galaxies, millions to billions of times more massive than the stellar mass black holes. As gas or stars fall in and are stretched and ripped apart by the enormous gravitational fields of these black holes, the resulting heated gas leads to correspondingly brighter electromagnetic flaring episodes, spread out over longer times, and recurring fitfully. These flaring episodes on the galactic scale have brightnesses which exceed thousands or tens of thousands of times the luminosity of the more peaceful steady-state emission produced by their stars or by the low and steady accretion of gas onto the black hole. Yet, bright as these electromagnetic galactic flares are, observations as well as simple physical arguments tell us that many of them must be accompanied by comparable or even larger outpours of energy in the form of cosmic rays, neutrinos and gravitational waves (Fig. 1.2).

Figure 1.2   A relativistic jet shooting out from the massive black hole at the center of the active galaxy M87, which is an incredibly energetic source of photons and particles.
Source: NASA Hubble Space Telescope.

1.3   The vast dark sea

The looming bulk of the dark Universe, alas, provides the greatest and least tractable mysteries. What are the dark energy and the dark matter, and what can we do to find out what they are, and how they operate?

Of these, dark matter appears to offer somewhat more promising or at least straightforward approaches for its investigation. For more than three decades, it has been studied indirectly through its gravitational effects on normal, visible matter. However, direct methods of investigation, such as capturing or analyzing the effects of dark matter interacting within laboratory detectors, appear at least possible as well. If the dark matter is not made up of hard-to-detect macroscopic objects, as seems to be the case after long and fruitless searches, it should consist of hard-to-detect elementary particles, for which there are some possible candidates. Those in the known arsenal of the Standard Model of particle physics, such as electromagnetic radiation at hard-to-detect frequencies, or neutrinos, appear to be ruled out. But there are many plausible extensions of the Standard Model which predict particles that could fit the bill, such as various types of weakly interacting massive particles (graced with the acronym WIMPS), or another type of hypothetical wimpy particle called axions. WIMPS are thought to be able to annihilate each other to produce neutrinos, which are in principle detectable with large neutrino detectors such as IceCube under the Antarctic ice or KM3NeT under the Mediterranean sea. In deep underground laboratories, WIMPS are also being searched for through the weak recoil they would impart to nuclei with which they (very rarely) interact. And one of the prime targets of large particle accelerators such as the new Large Hadron Collider (LHC) near Geneva, in Switzerland, is the detection of "something missing" when accounting for the energy budget of colliding high energy particles, which could indicate the creation of WIMPS. The latter, being weakly interacting, would leave the detector unnoticed, without paying their bill, so to speak, but leaving a noticeable gap in the collision energy balance.

Dark matter WIMPS can also annihilate by interacting with each other, leading to distinctive gamma-ray signatures which are being searched for with, among others, the recently launched Fermi Gamma-ray Space Telescope (formerly known as GLAST), and also with ground-based devices called imaging air Cherenkov telescopes (IACTs), such as HESS, VERITAS, MAGIC and CANGA-ROO. Besides their more spectacular and speculative task of probing the dark matter sector of the Universe, these space and ground instruments earn a hard living through honest, untiring and only slightly less spectacular studies of the more extreme forms of "normal" matter, such as black holes, gamma-ray bursts, supernovae, active galaxies, etc.

Dark energy is even harder to grasp, both experimentally and conceptually, than dark matter. The experimental study of dark energy is, for now, mainly confined to indirect methods. As in the case for dark matter, dark energy manifests itself most blatantly through its dynamical effects on the large scale behavior of the normal visible matter, in particular on the apparent acceleration of the expansion rate of the Universe. This is being studied by a variety of large scale optical surveys of distant objects, with new and proposed ground- and space-based experiments.

However, a theoretical understanding of the nature of dark energy, of what it is and how it fits in with the fundamental forces and other constituents of the Universe, remains perhaps the most challenging task of theoretical physics and astrophysics. If it is indeed a fundamental physical property, the answer is likely to lie at the interface between gravitation and quantum mechanics.

1.4   The great beyond

The study of both dark matter and dark energy pushes at the boundaries of particle physics and appears to require a unification of quantum mechanics and gravity, which is currently the most ambitious goal of theoretical physics. A major and very active component of this quest is the exploration of particle theories "beyond the Standard Model" (BSM). There are two major arenas where this is being played out. First, terrestrial experiments on very large particle accelerators such as the LHC or deep underground detectors such as Super-Kamiokande in Kamioka, Japan; experiments underway at Gran Sasso Laboratory in Italy and at the planned Deep Underground Science and Engineering Laboratory (DUSEL) in the USA, among others (Fig. 1.3). Second, theoretical models of processes in the very early Universe and related cosmological observations.

One critical epoch in the early history of the Universe is the so-called electroweak transition epoch, when the thermal energies of particles in the Universe had values comparable to those that are achievable in the LHC. There is also an even earlier epoch, during which an episode of greatly accelerated expansion is thought to have occurred. This is called the epoch of inflation, at a time when the Universe would have been so dense and hot that so-called Grand Unified Theories (GUTs) of particle physics hypothesize that three of the known forces of nature, the strong, the weak and the electromagnetic forces, would have been unified into a single interaction (e.g. [1]). And even earlier than that, at the so-called Planck epoch, the fourth force, gravity, would also have become comparable in strength to the other three forces, and the structure of space-time itself would have been a jumble of random quantum fluctuations. Somewhere in this imposing, chaotic landscape may lie the clues to unravel the nature of dark energy and its connection to the rest of physics, or at least that is the hope.

Figure 1.3 Aerial view of CERN, the European Center for Nuclear Research in Geneva, and the surrounding region. Three rings are visible, the largest of which (27 km in circumference) is the Large Hadron Collider (LHC). One of the goals of the LHC is the investigation of dark matter, within the broader context of physics beyond the Standard Model.
Source: CERN.

Another area where the microcosmos and the macrocosmos are intermeshed is the cross-fertilization between high energy physics and black hole astro-physics. One potentially interesting and exotic aspect of this arises in so-called low energy extra-dimensional theories (which are beyond the Standard Model, since they involve more dimensions than the usual four of space-time, e.g. [2]), where there is a possibility that proton collisions in the LHC at teraelectron-volt (TeV) energies could produce very small black holes. While the probability of this is acknowledged to be extremely low, even upper limits on it would provide useful constraints on possible non-standard models. Incidentally, it is worth noting that concerns that such microscopic laboratory black holes could pose a danger have been shown to be groundless [3, 4]. On a more abstract plane, black holes and particle physics mingle intimately in theories of quantum gravity. Both string theories and quantum loop gravity have made advances in describing the quantum properties of black holes, and have derived more or less self-consistent descriptions of black hole quantities such as the mass, spin, charge, information content, entropy, etc. [5–7]. However, these pursuits are still in their early stages, and the road ahead remains largely unfathomable.

It has also been suspected for a long time that black holes may play a role in the evolution of the early Universe. Some of the speculations include, for example, that black hole formation could lead to the currently observed photon-to-baryon ratio; that black holes could hide baryons which might otherwise have caused departures from the observed nuclear abundances of the chemical elements; that black holes might act as dark matter, or as a catalyst for nucleating galaxies, etc. Another speculation is that black holes could provide a feedback mechanism which, out of many possible Universes (the so-called multiverse [8]), selects the one where we happen to live [6]. And of course, the rate at which small and large black holes form in the more recent Universe, which is susceptible to direct observation, would provide a very powerful tool for tracing the dynamics and the evolution of star, galaxy and large scale structure formation. Ultra-high energy cosmic rays, neutrinos and gravitational waves, whether associated with these black holes, or perhaps other more exotic phenomena, will certainly provide unique probes to extend our current reach into the depths of the Universe.

1.5   The next steps

Mountaineers are familiar with the feeling of straining to climb a mountain range whose summit they can see and which apparently has only blue sky beyond, only to reach the presumed summit and discover that the view from there now opens new vistas of another, even higher mountain ridge. The process then repeats itself time after time, until (at least in earthly mountaineering) a final top is reached. The same is known from everyday hard work at an apparently impossibly large task; we know that the only way to accomplish it is to do it one step at a time, one day at a time, and just concentrate on the immediate task ahead, until we reach our goal.

What are some of the direction signposts and the first steps we can take towards these vast unknown territories of the Universe? Starting with the visible sector, the greatest challenges in the astrophysical arena are twofold: understanding the nature and dynamics of the expanding dark Universe, and unraveling the inner workings of its brightest concentrated high energy sources, such as supernovae, gamma-ray sources, super-massive black holes and their related objects. Due to their extreme brightness, which makes it possible to detect them out to the farthest reaches of the Universe, another crucial role of these sources and their messengers may be their acting as tracers of the development and dynamics of the Universe at the dawn of the stellar and galaxy formation epochs. Our horizons could be extended to even larger distances than now being reached if we were to detect from them ultra-high energy

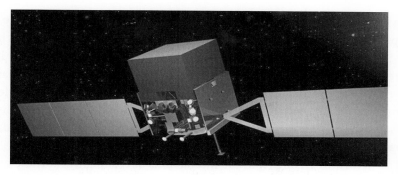

Figure 1.4   Artist's view of the Fermi Gamma Ray Space Telescope, launched in 2008, which is observing distant gamma-ray bursts, active galactic nuclei, pulsars and other objects, as well as providing limits on cosmic rays and setting constraints on dark matter models.
Source: NASA.

neutrinos resulting from ultra-high energy cosmic rays. Gravitational waves arising in these objects would also be able to reach us without any absorption from the largest distances, and these are the target of large gravitational wave observatories such as the Laser Interferometric Gravitational Wave Observatory (LIGO) in the USA, a similar observatory called VIRGO near Pisa in Italy, and a planned European spacecraft called the Laser Interferometer Space Antenna (LISA). Together with the more obvious visible tracers, these may help to track the "bulk" properties of the dark energy, as well as the details of the dark matter distribution (Fig. 1.4).

The most energetic type of radiation known so far, either from the laboratory or from the cosmos, are the ultra-high energy cosmic rays, and a major question is their possible relation to black holes, either massive or stellar. Are these cosmic rays astrophysical in origin, and related to active galactic nuclei, to gamma-ray bursts, or to supernovae? If so, they may shed light on the origin and nature of these objects. Or, alternatively, could they be the product of exotic processes beyond the Standard Model of particle physics in the early Universe? For their part, independently of any relation to ultra-high energy cosmic rays, the physics of black holes in active galactic nuclei and in stellar systems, gamma-ray bursts and supernovae involves extraordinary mass and energy densities which probe states of matter beyond anything which the laboratory can provide. And, as a population, they may play a very significant role in the development of large scale structure in the Universe.

Whatever their origin, at the enormous energies of 10^{20} eV the ultra-high energy cosmic rays surpass anything achievable in earthly accelerators, and provide an intimate link between the cosmological macrocosmos and the microscopic world of particle physics, at energies which may disclose features beyond the Standard Model of particle physics. This possibility remains even if, as it increasingly appears, they are not the product of the decay of exotic particles, but rather result from astrophysical acceleration in active galactic nuclei or in gamma-ray bursts. In all cases, the center of mass-energies in the collision of such cosmic rays with protons in the Earth's upper atmosphere is hundreds or thousands of times larger than the highest energies in the LHC.

The neutrinos arising from the interactions of cosmic rays at these energies also surpass by orders of magnitude any neutrino energies achievable in laboratories. Neutrino interactions, both at these terrestrially unachievable energies and at lower energies, are especially interesting, because neutrinos provide to date the only clear experimental evidence for physics beyond the Standard Model, through the phenomenon known as neutrino oscillations. This is related to the (non-Standard Model) phenomenon of the neutrinos having a small mass, which leads to neutrinos of different types changing identities as they travel over very large distances. The best known example of this is electron-type neutrinos from the interior of the Sun changing into muon-type neutrinos, as they make their way to the Earth. These "neutrino-flavor" changes and related phenomena are the subject of numerous laboratory, reactor, accelerator and underground experiments, using both terrestrially generated and cosmic neutrinos.

Such neutrino properties could have a direct bearing on the reason why the Universe consists mainly of matter (as opposed to anti-matter), instead of being a symmetric mixture of both. While the Universe may have started out with a uniform mixture, at some early point an imbalance must have set in leading to the survival mainly of matter, or baryons, a process called baryogenesis. Some of the leading theories attempting to address baryogenesis start out from leptogenesis, a process where leptons (which include neutrinos and other lighter particles such as electrons, etc.) become asymmetrical, which later through the weak interactions of baryons could lead to a baryon asymmetry.

2

The nuts and bolts of the Universe

2.1 The building blocks: elementary particles

2.1.1 Atoms and quanta

We are all familiar with the concept of atoms. We are made out of them, our surroundings are made out of them, and our Universe is made out of them. The name derives from the Greek, meaning "indivisible", which conveys the idea that these are the smallest building blocks out of which the Universe is built. In the early 1900s the smallest units were indeed considered to be the atoms, consisting of a central more massive kernel, the nucleus, surrounded by a cloud of orbiting, much smaller and lighter particles called the electrons. The electrons were found to have negative electrical charge, while the much heavier nucleus had an equal amount of positive electrical charge, which was attributed to heavy particles called protons. Later, in the early 1930s, it was found that the nucleus contained other particles as well, slightly heavier than the protons but electrically neutral, which were consequently given the name of neutrons.

For a while these appeared to be all of the basic building blocks of matter. Different atoms, such as hydrogen, helium, carbon, iron, etc., consisted of a nucleus which differed by containing increasing amounts of protons, and except for hydrogen, a comparable or slightly larger number of neutrons, and around the nucleus a number of electrons matching the number of protons, so as to ensure electrical neutrality. This was thought to be what ordinary matter consists of, and in fact this picture continues to be basically correct to this day, except for the fact that it is not the complete picture. First, the nuclear particles have since turned out not to be elementary at all but to have sub-constituents, and second, a new theory had to be developed to correctly describe the mechanics of the atomic and sub-nuclear world, which differed greatly from the old Newtonian mechanics describing the classical world of planets, pulleys, inclined planes, cars, etc.

This new theory of atomic and sub-nuclear physics is called quantum mechanics, where the word "quantum" means that the quantities involved come in discrete chunks, or quanta. The energy, the impulse, the angular momentum and most of the other properties of the electrons, protons, etc. are "quantized", i.e. they come in discrete multiples of a small number. Previously, in classical Newtonian mechanics and Maxwellian electrodynamics, it was thought that the various physical quantities associated with a system, such as its energy, momentum, etc., could adopt any of a continuum of possible values. There was no obvious reason why any possible value could not be mentally halved and give an equally possible value. The need for a discretization of physical quantities originated with Max Planck in 1900, who showed that electromagnetic radiation had to be quantized, i.e., it did not consist of continuous infinite waves of arbitrary frequencies but of discrete "wave packets" or "photons", carrying a discrete amount of energy given by the product of the wave frequency times a small constant number now denoted $\hbar$ (h-bar) and known as Planck's constant.

The concept of quanta was extended to material particles by Einstein and later to atoms by Bohr, in the first two decades of the 1900s, and quantum mechanics in its basic current form was laid down in the mid-1920s by Heisenberg, Schrödinger and Dirac. In quantum mechanics, all dynamical quantities are discrete multiples of some smallest unit involving Planck's constant $\hbar \simeq 10^{-27}$ erg s. These discrete quantities characterizing the particles are called the *quantum numbers*. Quantum mechanics differs from ordinary mechanics also in that it deals not with deterministic predictions of the future position and the dynamical quantities, but with the probabilities of being at some later time at some position with some particular values of the quantum numbers. One aspect of this is that we cannot determine all the relevant variables of a particle with high precision. For instance, if the position x of a particle is measured to within an error Δx, its momentum p cannot be determined to an accuracy better than $\Delta p > \hbar/\Delta x$, i.e., the uncertainties in the two quantities satisfy in general the relation

$$\Delta x . \Delta p \geq \hbar, \tag{2.1}$$

which is a statement of the Heisenberg Uncertainty Principle.

Another development around this time was the realization that all particles have a spin, which can be thought of as the particles spinning about some axis like a top, or like a tennis ball, in addition to their motion through space. According to quantum mechanics and also to experiments, the amount of spin comes in integer or half-integer multiples of Planck's constant $\hbar$. Thus, protons and electrons have, in quantum mechanics, a probability density describing an orbital motion, somewhat like the Earth around the Sun, and describing also their spin, somewhat like the Earth and the Sun spin around their own axes. All particles in quantum mechanics can have a spin, just like a thrown tennis ball or a football can be imparted a spin. According to experiments and their quantum interpretation, particles like the electron, the proton and the neutron have a half-integer spin, which means that its value is $(1/2)\hbar$, and the spin can be either right-handed or left-handed along the direction of motion. Other particles, such as photons, however, have an integer spin; in the specific case of photons this is $\hbar$, while there are other particles whose spin is $2\hbar$, $3\hbar$, etc. The spin is another way of describing the polarization of the electromagnetic waves (e.g., as seen through polarized sunglasses).

Interestingly, when describing the statistical properties of particles of half-integer or integer spin, it is found that they obey different statistical laws [9]. That is, when describing the probabilities of finding x amount of a certain type of particle at a certain location with certain sets of quantum numbers, these probabilities are drastically different for half-integer or integer spin particles. Half-integer spin particles cannot be at the same location and have the exact same quantum numbers (energy, spin, etc.). This is an experimental fact, also called Pauli's Exclusion Principle, and the type of statistics obeyed by such half-integer spin particles is called Fermi–Dirac statistics. This is very important, as we will see later, and such half-integer spin particles are called *fermions*. Most of the known massive particles, such as protons, neutrons, electrons, etc., are fermions. On the other hand, integer spin particles, such as photons, obey a different type of statistics, called Bose–Einstein statistics, and for this reason integer spin particles in general are called *bosons*. Unlike fermions, bosons can coexist in the same location with the same quantum numbers in any amounts. Unlike fermions, which may be considered individualistic or stand-offish, bosons may be termed gregarious. This is what makes possible devices such as the laser, where a great many photons of exactly the same frequency and polarization can bunch up together, thus greatly multiplying their collective effects.

2.1.2 *Anti-matter, neutrinos and the particle explosion*

Starting in the early 1930s, it was found that besides ordinary matter there existed other types of matter, far from ordinary. For several decades it had been known that cosmic rays, which are mainly charged particles such as electrons, protons and heavier nuclei, arrived at the top of the Earth's atmosphere from outer space with extremely large energies. When these interacted with a detector they produced secondary particles, among which were found particles with the same mass as electrons but with an electric charge of the same value but opposite, positive sign. Such anti-electrons, or positrons, had been predicted theoretically by Dirac in the late 1920s, and this was the first example of what has come to be known as anti-matter.

Also in the 1930s another new type of particle, even more mysterious, made itself increasingly more evident. These particles occurred in some nuclear reactions and radioactive decays, and appeared extremely hard to detect directly. However, their presence became increasingly obvious due to the fact that in the nuclear reactions, when measuring the energy and the momentum of the initial and final particles, which are thought to be subject to an overall conservation law, the accounting fell short, unless one postulated the existence of such undetected particles. They had to have zero electric charge, otherwise they would have been easier to detect, and they had to be either massless or have extremely small masses. These particles, whose existence was first postulated by Pauli, were given the name of *neutrinos* by Fermi [10].

During the 1940s and 1950s other new, very short-lived particles were found in cosmic-ray interactions as well as in particle collisions produced in laboratory accelerators. These were heavier than the electron but lighter than the proton, with names such as pion, muon, etc., some being negatively or positively charged, while others were neutral. Other types of anti-matter started being found as well, such as anti-protons (labeled $\bar{p}$), which have the same mass and other properties as the usual protons, but with a negative electric charge. However, anti-matter was found to be extremely short-lived in the presence of ordinary matter, since the anti-particle quickly annihilates itself with one of its ubiquitous (ordinary matter) partners, emitting two photons. By the late 1950s and 1960s, unstable particles and anti-particles even heavier than protons and neutrons were being found in increasing numbers, in what came to be called the particle zoo. Being unstable, all of these exotic particles decayed in a very short time into other, more normal, stable particles.

Some of the more common particles and their properties are listed in Table 2.1. The masses are measured in energy units of megaelectronvolts (MeV) (divided by the speed of light squared). This is because energies are easier to measure in particle physics, and the mass follows from the well-known $E = mc^2$ relation. The MeV is a natural energy unit in nuclear physics, but it is extremely small compared to everyday quantities. For example, one calorie is equivalent to 2.6×10^{13} (26 trillion!) MeV, and an average human eats a few thousand calories per day, which is about 5×10^{16} MeV (fifty thousand million million megaelectronvolts! For some other common units and their equivalents, see Table A.1 in

Table 2.1. *Properties of some of the more common particles*

Type	Name	Symbol	Mass (MeV/c^2)	Mean life (s)
Baryons	proton (anti-proton)	$p, \bar{p}$	938.2773	$\gtrsim 10^{32}$ year
	neutron	n	939.5656	887
Mesons	pion (charged)	$\pi^{\pm}$	139.57	2.6×10^{-8}
	pion (neutral)	π^0	134.98	8.4×10^{-17}
Leptons	electron (positron)	$e^{\pm}$	0.511	stable
	muon	$\mu^{\pm}$	105.658	2.197×10^{-6}

the Glossary). The mean lifetimes of the particles, when they are unstable, are indicated in seconds.

For a long time it was thought that all of these particles, both those making up the ordinary stable matter and the exotic unstable ones, were "elementary" particles. That is, particles which have no sub-units, they are just themselves, period, the only qualifiers being their quantum numbers. The problem was that there were so many particles that any sort of classification and categorization of properties which could lead to a comprehensive theory was extremely difficult, and indeed frustrating.

2.1.3   Elementary, dear Watson

In the late 1960s and early 1970s, however, it was realized that protons and neutrons, and indeed many of the unstable particles arising in high energy collisions, were not elementary after all. They turned out to be made up of sub-units which came to be known as *quarks*, most being made up of different combinations of the two commonest quarks, called the "up" and "down" quarks [9]. However, electrons and positrons, as well as photons, still remain as elementary particles, with no known sub-structure. The electrons have unit (negative) electric charge (the unit is labeled e), whereas the quarks have fractional electric charges, the up quarks having $+(2/3)e$ and the down quarks having $-(1/3)e$. Protons, neutrons and most of the heavier unstable particles (collectively labeled *baryons*, meaning heavy) consist of three quarks, in combinations such that their total charge gives the observed electrical charge. That is, the proton is a combination *uud*, of charge $(+2/3 + 2/3 - 1/3)e = +e$, while the neutron is a combination *udd*, of charge $(+2/3 - 1/3 - 1/3)e = 0$. The quarks are of course fermions with half-integer spin, their combination giving the resulting spin of the protons and the neutron. Also, being charged, a quark q has a corresponding anti-quark (labeled $\bar{q}$) which has the opposite charge sign, so that anti-protons are made up of anti-quarks, etc.

Table 2.2. *The elementary fermions*

Sector	1st family	2nd family	3rd family	$Q/|e|$
Leptons	e	μ	τ	-1
	ν_e	ν_μ	ν_τ	0
Quarks	u	c	t	$+2/3$
	d	s	b	$-1/3$

Other combinations of quarks give rise to most of the various unstable particles which are encountered for brief times in high energy collisions. There are two such groups of unstable particles consisting of quarks [11]. One group consists of medium-weight particles (compared to the proton), called *mesons* (from the Greek word for "middle"), which are made up of a quark and an anti-quark. These include particles with names such as the pion, the K-meson, the D-meson, etc. The other group contains the aforementioned *baryons*, heavier than the mesons, which consist of combinations of three quarks, and includes the protons and neutrons, as well as large numbers of different unstable particles heavier than the proton. Most of these unstable particles, aside from the pions, however, include two additional families of quarks, besides the first family of up and down quarks which make up the ordinary stable matter. The second quark family consists of the *strange* (s) and *charm* (c) quarks, which are heavier than the u and d, and the third family consists of the *bottom* (b) and *top* (t) quarks, which are even heavier than the others (the t is a hefty 180 times the mass of the proton).

Not all unstable particles are, however, made up of quarks. The *leptons* are another group of elementary particles, which share with the quarks the property of being fermions, but which are lighter than the quarks, mesons and baryons. The leptons consist of elementary particles, without sub-structure, some of which are stable (such as the electrons) while others are not. The leptons are again divided into three families, or *flavors*, coming in pairs consisting of one electrically charged and one neutral fermion. The first flavor or family consists of the electron e^- and the electron neutrino ν_e, the second flavor consists of the muon μ^- and the ν_μ, and the third flavor consists of the tauon τ^- and the ν_τ. These also have their corresponding anti-particles (e^+, $\bar{\nu}_e$, μ^+, $\bar{\nu}_\mu$, etc.). These families parallel the quark families. The whole set of elementary fermions is shown in Table 2.2, where Q indicates the electrical charge in units of the elementary charge e.

These particles interact through various types of forces, the interaction occurring through the exchange of an intermediary particle which is a boson, more technically called a gauge boson. These forces and the corresponding exchange bosons are described in the next section.

2.2   The forces: three easy pieces and a harder one

Four basic types of forces, or types of interactions, are known so far in Nature. These are the electromagnetic, the gravitational, the weak and the strong interactions. The effects of the first two are felt in everyday life, while the second two appear mainly in nuclear and particle physics processes. These forces appear to emanate from individual sources (masses, charges, etc.) which are either particles or are made up of particles. In the case of electromagnetism and gravity, these forces are most readily apparent from large macroscopic amounts of matter, but electromagnetism plays a significant role also when considering the smallest indivisible amounts of matter, elementary particles such as electrons and quarks. In the case of the weak and the strong forces, these first become apparent when considering elementary particles, or small groups of them, although large amounts of them can transcend the sub-nuclear realm and lead to wondrous large scale manifestations, such as nuclear reactors, explosions and stellar energy generation. Whereas electromagnetism and gravity "in bulk" have very good classical (macroscopic) descriptions given by classical mechanics and Maxwellian electrodynamics, similar "bulk" descriptions are not adequate in the case of the weak and the strong interactions. The latter two can only be described adequately by means of a new type of description, quantum mechanics, which as mentioned is based on the postulate that all physical quantities associated with elementary particles come in small discrete chunks, or quanta. Electromagnetism has, in addition to a successful macroscopic Maxwellian description, also a quantum description, called quantum electrodynamics, which is important in the atomic and nuclear world. For gravity, however, the search for a quantum extension of the macroscopic theory is still on.

To each of these four forces there are, in the language of quantum mechanics, associated messenger particles, which mediate the interaction between the sources susceptible to that particular interaction. These messengers act like springs between masses, or like balls bouncing back and forth between the sources, transferring energy and momentum between them. These messenger particles are themselves also quanta, with discrete energies, momenta, etc., which under their technical name are called gauge bosons. While the "sources", that is the particles which interact through the forces are fermions, the messengers carrying the force between them are bosons. The names of these messenger bosons for the electromagnetic, the gravitational, the weak and the strong forces are, respectively, the photon, the graviton, the W and Z bosons, and the gluon [11, 12]. We discuss these four forces and their messenger bosons in turn.

2.2.1   The electromagnetic force

While on the large astronomical scales of planets, stars, galaxies, etc. the dominant force is gravity, almost all the information we have about these objects comes to us through their light [13]. Classically, light is a form of electromagnetic radiation, which in its quantum description comes in discrete quanta called photons. When we observe galaxies, or anything else for that matter, we do it by collecting myriads of photons, which enter our eyes or our telescopes, and are analyzed there by various physiological or electronic devices. The basic sources of these electromagnetic quanta are elementary particles endowed with electric charge, such as electrons, quarks, or smaller groups of them. The electromagnetic force is attractive between electrical charges of opposite sign, and repulsive between charges of the same sign.

The photons are in fact the messenger, or gauge bosons, which mediate the electromagnetic interactions between charged elementary particles, and they are described by a few basic quantities, such as their energy (or their frequency or wavelength, which are also present in the classical description), their momentum, and their spin or polarization. Photons, however, do not have any mass, as far as we know, and they travel (in vacuum) at the speed of light, $c = 300\,000\,\text{km s}^{-1}$, which according to special relativity is the maximum physical speed achievable by any object.[1] The fact that photons transmit an electromagnetic force can be appreciated also at the mundane level, by the fact that sunlight impinging on our skin causes a sensation of heat. This results from the photons giving energy to electrons and molecules in our skin, whose energy of motion is dissipated and absorbed, resulting in a sensation of warmth. The electromagnetic waves, consisting of many photons traveling together and behaving similarly, can be described through Maxwell's equations as a traveling set of forces exerted in a direction perpendicular to the direction of travel. These forces reverse their sense (say from left to right, or from up to down) at regular intervals in space (if observed at a given time) or with a certain frequency (if observed at a fixed point in space). These alternately reversing forces act on electrically charged particles, such as electrons, and make them wiggle in response (Fig. 2.1). A more macroscopic manifestation of the electromagnetic force is

[1] The fact that the speed of light cannot be exceeded by any kind of physical object or signal, a fact amply verified by experiment, is the basis of the special and the general theory of relativity, which is an integral part of both Maxwellian and quantum electrodynamics.

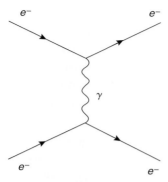

Figure 2.1   Electromagnetic interaction between two electrons, mediated by the electromagnetic gauge boson (the photon, represented by the wavy line). This figure, an example of what are called Feynman diagrams, represents electron–electron scattering via photon exchange. Time increases to the right.

illustrated by the reverse effect, when electrical currents, that is bunches of electrical charges, are made to circulate in a circuit around magnets such as in electrical motors, resulting in the bodily motion of the rotor which energizes diverse types of machinery. This is because the electric and magnetic fields of force are intimately linked and act on each other, hence the unified name of electromagnetic force.

Photons or electromagnetic waves whose wavelengths are of order $\lambda \sim 10^{-5}$ cm are called optical photons (or light), these being the photons to which our eye is sensitive. At longer wavelengths, the electromagnetic radiation consists of, successively, infrared, sub-millimeter and radio photons, while at shorter wavelengths we have ultraviolet, X-ray and gamma-ray photons. The Sun emits most of its electromagnetic energy in the form of optical photons (that is why our eyes developed to be sensitive to optical photons), but it also emits smaller fractions of energy at practically all other wavelengths. However, other types of cosmic sources are found which emit most of their electromagnetic energy, or are primarily detected, at different wavelengths, such as gamma-ray burst sources, X-ray pulsars, or radio-galaxies.

The electromagnetic force is the best understood force in nature, and it plays a major role in everyday life, from controlling molecular structures in our and in other bodies, animate and inanimate, to being the basis of countless industrial applications such as motors, lighting, radio, television, telephony, wireless, etc. An important property of this force is that it is long-range: the electric field of a single charge, that is the force experienced by another electric charge located at a distance r away from the first charge, falls off as the inverse square of the distance, $F_{EM} \propto 1/r^2$. This is what makes radio and other electromagnetic signals, consisting of individual photons, propagate not just from some station to our home, but also over astronomical distances. It is thanks to this long-range property that almost all of what we know about the Universe has been learned through analyzing the photons emitted by various astronomical objects.

An important factor in the electromagnetic interactions is that there are two types of electrical charges, positive and negative. The forces binding the electrons to the nucleus in atoms are electromagnetic in nature, and so are the intermolecular forces. However, in molecules there are so many (negatively charged) electrons and so many (positively charged) protons that at some small distance away from the molecule the two signs of the charges cancel out, making the electromagnetic interaction between molecules become effectively a short-range one. This phenomenon is called shielding. Shielding is however not perfect, and it is the residual electromagnetic force which keeps together the molecules of fluids and solids, or the molecules in our body as well as the molecules in a wall, which we can push with our hand without one penetrating the other.

In the 20th century, the 19th-century classical "macroscopic" description of Maxwellian electromagnetism was successfully translated into the language of quantum mechanics. This is the theory of quantum electrodynamics which takes fully into account the fact that the electromagnetic field consists of individual photons, the quanta of this field, and that these interact with particles whose properties are also quantized and obey quantum mechanics. One of the great successes of quantum electrodynamics, due to Dirac, was the prediction of the existence of anti-matter, or anti-particles. The quantization of the electromagnetic field, besides providing a far deeper understanding of the basic nature of this interaction, has had an enormous practical impact on various industrial applications, such as lasers, optical fiber communications, data encryption and quantum computing, etc., which in turn have greatly impacted the development of detectors for astronomical as well as laboratory measurements.

2.2.2   The weak force

The weak and the strong forces occur in nuclear physics and in high energy interactions between elementary particles, such as in large laboratory accelerators, in stars or in cosmic rays accelerated by cosmic sources. In contrast to the electromagnetic and the gravitational forces, the weak and the strong forces are felt only at short range, over dimensions comparable to the sizes of nuclei and elementary particles. Also, in contrast to electromagnetism and gravity, there are no "classical" or macroscopic descriptions of these nuclear forces. Quantum mechanics is needed to describe them even at the simplest level, when individual nuclei or particles are considered, or small assemblies of them.[2]

The quantum mechanical description of the weak force is modeled after quantum electrodynamics. The latter is a theory which has been fantastically successful, allowing incredibly precise calculations which agree with experiment to within 10 digits and more accuracy. The weak interaction has, after electromagnetism, the next best developed quantum theory, although the level of complexity is significantly higher and the level of understanding is much more approximate. In its modern form the weak interactions have in fact come to be described in a completely similar manner as electromagnetism, in a joint quantum formulation called the electroweak theory. In this joint theory, these two interactions and their experimental phenomenology differ substantially from each other at energies below the so-called *electroweak* energy scale, which is about 100 GeV, but above this energy the two sets of phenomena start to become increasingly similar. At energies somewhat below the electroweak scale this has been verified experimentally, and a study of these phenomena at the electroweak scale and above is one of the major goals of modern accelerators such as the LHC at CERN in Geneva.

The weak interactions were first observed in radioactive nuclear decays, and more generally they involve elementary particles such as leptons and the quarks making up nucleons or other unstable particles. They are characterized by always involving neutrinos, which as mentioned are extremely light, electrically neutral elementary particles. According to the Standard Model of particle physics, neutrinos would actually be massless, and consequently they would be expected to travel at the speed of light in vacuum. But one of the reasons why we know that physics beyond the Standard Model is needed is that now we know that neutrinos do have a very small mass, as discussed below. Also unlike the photon, of which there is only one kind, there are three kinds or "flavors" of neutrinos: the electron neutrino, the muon neutrino and the tau neutrino, which participate in different types of weak processes. The sources or particles producing the weak interaction are endowed with a "weak" charge, which is related to their electrical charge.

There are two major reasons why these interactions are called "weak". One of them is that the neutrino, which is characteristic of such interactions, is extremely hard to detect, unlike the photon – the neutrinos interact extremely

---

[2] However, when very large numbers of particles are considered, a suitable averaging of the quantum mechanical equations leads to the usual macroscopic classical description of matter in bulk.

weakly with any detector. The interaction rate is so minuscule that more than 5×10^{13} (50 trillion) neutrinos emanating from nuclear reactions in the Sun pass through our bodies every second, without causing any harm. Unlike the photons from the Sun, they don't stir up the electrons in our body molecules, they just go right through them (except so rarely as not to make any difference). The other main reason why their name is appropriate is that the weak interactions occur extremely slowly. For instance, the chain of nuclear reactions in the interior of the Sun, which generate the energy (and the photons) ultimately giving rise to life on Earth, involve both weak and strong processes, but it is the weak interactions which take the longest to occur. They set the slow pace of evolution of the Sun, and in fact if they had been any faster, biological evolution and life on Earth would not have had the billions of years necessary to reach its current state.

The messenger particles of the weak interactions are of three types: the W^+, W^- and Z^0 bosons. These, unlike the photons, are massive particles – in fact, quite massive, about 80 and 90 times heavier than protons. The W bosons are endowed with electrical charges indicated by the $+, -$ superscripts, while the Z bosons are electrically neutral. The W and Z bosons mediate between particles carrying a weak charge, just as the photons mediate between particles carrying electrical charges. The fact that the messenger particles are so heavy is the basic reason why the interaction is of short range. The messengers are so heavy and sluggish that they can't travel very far, unlike the massless, infinitely nimble photons.

One of the aspects of the unified electroweak theory is that at energies above that of the W and Z boson mass-energy (which is roughly the "electroweak" energy) it considers the weak bosons as massless, just as the photons are (the latter are massless however at lower energies as well). Below the electroweak scale, however, the weak bosons acquire a mass. This is part of the more general theory of the Standard Model, which generates the mass of these and other particles below the electroweak scale through the intermediary of a new complex quantum field, called the Higgs field. This field behaves as a scalar (instead of as a vector, such as the electric field), and it has the property of allowing at high energies a description of the electroweak theory where all particles are massless (fermions and bosons), while below the electroweak energy the fermions and some of the bosons acquire masses, while leaving the photons massless and predicting the existence of a massive scalar particle called the Higgs boson. The electroweak theory has had numerous successes, such as predicting the mass of the W and Z bosons, and explaining various other aspects of the weak interactions. This success has motivated the consideration of other types of scalar fields at high energies, such as those invoked to explain inflation and dark energy (discussed in Chapter 3). The mechanism for generating the masses of the particles is called the Higgs mechanism, and the mass of the predicted Higgs boson is expected to be in the range of $\gtrsim$ 100 GeV. Discovering the Higgs particle is one of the prime targets of the LHC and similar machines.

2.2.3   The strong force

The strong force is, together with the weak force, the other type of interaction which acts only over a limited range of distances, of the order of the size of nuclei or smaller. As in the case of the weak interactions, the strong interactions can only be described with any degree of success in a quantum formulation, the modern version of which is called quantum chromodynamics (QCD). As the name implies, the forces binding the quarks into nucleons (protons and neutrons) and binding the nucleons inside the nuclei are extremely strong. This enormous strength is what causes the splitting of a nucleus (fission) or the creation of more complex nuclei (fusion) to release the huge amounts of energy locked inside the nuclei in nuclear bombs and in nuclear reactors.[3] As a rough comparison, the fission of one kilogram (kg) of fissile material can deliver, undergoing strong nuclear reactions, an energy comparable to that which 1 kilotons (one million kilograms) of TNT would deliver through chemical reactions. Thus nuclear reactions are roughly a million times more efficient at delivering energy than the most energetic chemical reactions, which essentially depend on electromagnetic interactions.

The strong force acts between nucleons in nuclei, or rather between the quarks that make up the nucleons or other unstable particles. The quarks are the sources of the strong force, and as mentioned there are six types of quarks. Of these, the up and down quarks are the most common ones, making up the stable nucleons, the proton and the neutron. The other four types of quarks, the strange, charm, bottom and top quarks, are heavier and appear in the much rarer fleeting particles produced in very high energy particle collisions, and the six quarks are arranged in three families, or generations (Fig. 2.2). Each of the quarks in each family is endowed with three possible types of strong charge, called "color" charges, hence the name of quantum chromodynamics for the theory describing them. These colors are generally called red, blue and green (r, b, g). All particles made up of quarks, which are subject to the strong force, are posited to be color blind, or color neutral, i.e., they have quarks whose colors neutralize each other. The combination of r, b and g is neutral. There are also

---

[3] This might at first sight seem at odds with the short-range nature of the strong force. The long-range macroscopic effects of the strong nuclear force can be, and are, produced by bringing close together large amounts of strongly interacting particles so that their collective energy generation irradiates and heats up the neighboring matter, leading to electrical currents or large scale shock waves.

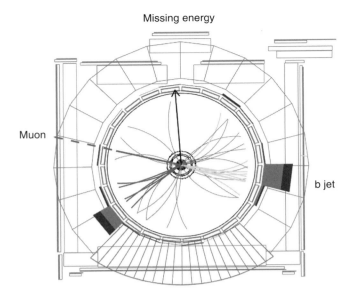

Figure 2.2   A particle collision event display in the CDF detector at FermiLab showing a single top quark event. Such collisions lead to jets of quarks and gluons, here showing two jets plus a neutrino ("missing energy") track and a track in the direction of a muon escaping from the decay of a top quark.
Source: Courtesy of the CDF Collaboration.

anti-colors ($\bar{r}$, $\bar{b}$, $\bar{g}$) for the anti-particles, the sum of which is also neutral. Other unstable particles made up of only two quarks, i.e., mesons such as the pion, must consist of quarks with one color and the same anti-color.

The messenger particles or gauge bosons mediating the strong color forces are the *gluons*. The gluons are massless and electrically neutral. Since they have to mediate between six different types of quarks with three different colors, there are eight different kinds of gluons, each of which carries a color charge and a different anti-color charge, to ensure that the particles between which they mediate remain color neutral after the interaction. The fact that the gluons, i.e., the messenger particles, carry colors means that they themselves can act as color charges, i.e., they are subject to strong interactions among themselves. This is unique to the strong force: none of the other three interactions have messengers carrying the charge corresponding to the interaction, only the sources do. This means that quantum chromodynamics is a more complicated theory in this respect as well: not only is the number of charges larger, but the messengers can interact among themselves.

The fact that the gluons are massless might suggest that the range of the interaction is infinite, as in the case of the photons. However, since there are

Table 2.3. *Approximate comparison of the
relative strengths of the four basic interactions*

Strong	Electromagnetic	Weak	Gravitational
1	10^{-2}	10^{-7}	10^{-39}

three types of color charges and since the particles must be color neutral, this leads to a cancellation of the color forces beyond nuclear distances, making the strong force essentially short range. The fact that the gluons can interact with themselves, unlike the photons in electromagnetism, leads to a phenomenon of anti-shielding of the color charges, which has important implications for the dynamics and kinematics of particle interactions at very high energies [1]. It appears rather complicated, but it all works out, and quantum chromodynamics is a very successful theory, which has allowed much progress to be made in the understanding of the strong interactions and particle physics in general.

2.2.4   *The gravitational force*

The gravitational force is by far the weakest of the four forces, much weaker than the nuclear "weak" force. If one compares the forces due to the four types of interactions between two particles of equal masses and charges across the same distance, the relative strengths are shown in Table 2.3.

Despite its extreme weakness, gravity is the most obvious of all forces: we see apples falling, we feel the weight of heavy objects, etc., and this is because *across moderate to large distances* gravity can overwhelm the other forces. One reason for this is that it shares with electromagnetism the property of being a long-range force, which also drops off as the inverse square of the distance, $F_G \propto 1/r^2$. It is because of this long-range property that the effects of these two forces can be felt over macroscopic distances. This makes their effects more palpable to the human senses, so they have been known and studied from much earlier times than the nuclear strong and weak forces, which are microscopic-scale short-range effects, requiring special instrumentation for their study. The gravitational force is the one that has been known and studied for the longest time, and in its classical Galilean and Newtonian form is perhaps the best understood. The motions of the planets, the fact that our bodies are "weighted down" by the gravitational attraction of the Earth, etc., are phenomena which can be appreciated with the eyes and with the senses, even without instruments, and this study underwent enormous development from the 17th through the 19th centuries.

However, the all-pervading obtrusiveness of the gravitational interaction relies on an additional important reason. This is that, unlike the electromagnetic interaction, it has only one type, or sign, of gravitational "charge", namely the mass. There are no negative or positive masses, just masses. This means that they cannot cancel each other out at a large distance. All masses produce attractive forces proportional to their mass, falling off with distance as $1/r^2$, no matter what other matter is between the source and the point of observation. And because there is only one sign of the mass, there is no screening of the gravitational force (whereas screening of the electromagnetic force can reduce the latter, in bulk matter, to an effectively short-range force). This is the reason why our bodies are attracted to the Earth (our "weight") by the gravitational force, since the other three forces, although microscopically stronger than gravity, have their effects canceled out by virtue of their effective ranges being much smaller than our body size. In fact, gravity can "coop up" the other interactions (Fig. 2.3).

The understanding of gravity underwent a major qualitative jump starting in the early 20th century. This was triggered by the fact that when one looks at the finer details of the motion of planets very near the Sun (Fig. 2.3), or when one considers very large masses such as contained in large expanses of the Universe, Newtonian mechanics leads to small but noticeable inconsistencies (and this gets worse, nowadays, with extremely dense and compact objects such as neutron stars or black holes). To treat these phenomena, one has to use a more complicated description, based on Einstein's general relativity. In this latter form, gravity is not even described as a force anymore, but rather as a distortion of space-time which results in the observed dynamics of the massive bodies being considered. Thus, for instance, the larger mass of the Earth distorts the structure of the space-time around it, and the Moon simply follows, or freely falls along, the natural curvature of this space-time in which it finds itself, resulting in its motion around the Earth. This description of celestial mechanics is somewhat harder to grasp, but as long as one is considering large macroscopic masses it provides a rather detailed and accurate mathematical machinery to describe the behavior of matter. Gravitation, in this description, still acts on the macroscopic scales, and one does not need to consider discrete chunks or quanta of the gravitational field: its bulk properties describe essentially everything that one observes macroscopically. General relativity even describes the large-scale gravitational equivalent of Maxwell's electromagnetic waves, namely gravitational waves. These are ripples in space-time, which travel at the speed of light, and which are being searched for with large experiments such as LIGO and VIRGO (discussed in Chapter 9).

However, even with this version of (general relativistic) gravity, conceptual problems arise if we consider extremely large mass or energy densities, that is,

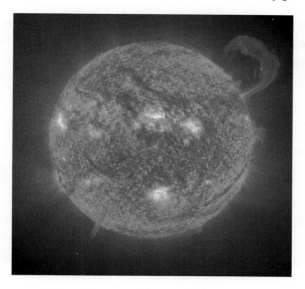

Figure 2.3 Our Sun, as seen by the SOHO satellite in ultraviolet light. The Sun is the source of all life on Earth, thanks to the light and heat that it delivers. The gravitational, electromagnetic, weak and strong forces all play important, concerted roles in the Sun, which is the ultimate environmentally friendly nuclear reactor in our cosmic backyard. The nuclear reactions in its core produce copious neutrinos, which escape but do not harm us; they produce nuclear waste, which is trapped by gravity in the core; and they produce gamma-rays, which multiply and lose energy as they slowly diffuse out, until they emerge mainly as beneficial optical sunlight.
Source: NASA.

extremely large amounts of matter or energy compressed into extremely small regions of space-time. In this case quantum effects are expected to become important, and the quanta of the gravitational field, the *gravitons*, need to be considered as separate discrete entities. This regime is encountered near the central "singularity" which would appear close to the geometrical center of black holes, or in the very early Universe, very close to the instant described as the Big Bang. This would require a quantum theory of gravity, which however is so far non-existent. There are candidate theories which have the elements of a quantum gravity theory, such as string theory, or quantum loop gravity, which attack this problem and absorb huge amounts of effort by some of the sharpest researchers, but so far with only suggestive results pointing towards an ultimately workable theory [7].

Thus, despite being the oldest and perhaps the best understood force in its macroscopic form, in its wished-for microscopic quantum formulation gravity remains the most recalcitrant among the four forces. The chamber concert of modern physics consists, at the moment, of three somewhat easier pieces, and a fourth one which apparently is simpler but upon closer inspection turns out to be more puzzling than the other three.

2.3   Beyond the Standard Model

The above-mentioned incompleteness of the gravitational theory is just one of the signs that some important pieces are missing from the jigsaw puzzle.

Another sign is that neutrinos, which are key participants in the weak interactions, were for a long time happily considered to be massless, even in the unified electroweak theory. However, according to experiments in the last one and a half decades, it appears that neutrinos, unlike what is postulated in the Standard Model of particle physics, do have very small masses. This means that they must travel at speeds extremely close but not quite equal to the speed of light. Their very slight sluggishness is caused by their tiny mass, which slows them down ever so slightly. This mass is also tied to the fact that neutrinos of different flavors can, as they propagate, switch from one flavor to another ("oscillate" between flavors). These are phenomena which are definitely beyond the Standard Model, and the study of such BSM phenomena is one of the major frontier areas of physics. These also have interesting astrophysical implications, which are discussed in some of the subsequent chapters.

Yet another indication that skeletons remain lurking in the closet is that the electroweak theory can explain the masses of fermions and bosons, but it requires a large number of ad-hoc parameters to do so, including the coupling strengths [14].

Looking ahead, theorists guess that if the electromagnetic and the weak forces become united at energies $\gtrsim 100\,\text{GeV}$, at even larger energies one would expect that the strong force should also become unified to the other two. There is in fact experimental evidence indicating that the strengths of the electromagnetic, the weak and the strong forces tend towards a convergence at energies of order $10^{16}\,\text{GeV}$, an energy which is however well beyond the reach of even planned accelerators. The search for such "Grand Unified Theories" (GUT, for short) is a major field of ongoing activity [1].

A major group of such GUT theories is based on a new type of symmetry between particles called supersymmetry (abbreviated SUSY GUTs). This considers the possibility of bosons and fermions inter-converting, and posits the existence of "superpartners" for each particle. To each fermion corresponds a boson, given the name "sfermion", and to each boson corresponds a "bosino" superpartner. For example, each "quark" (a fermion) has a "squark" superpartner, which is a boson; while the Z-particle (a boson) has a "zino" superpartner which is a fermion.

While electroweak theory, GUTs, SUSY GUTs, etc. were originally motivated by laboratory experiments and particle physics theory, these ideas soon spilled over into cosmology. In the very early Universe, of course, the Big Bang model predicts energy densities which are so high as to exceed anything in the laboratory, providing a likely arena where these ideas can play themselves out. One such scenario involving scalar fields modeled after the Higgs field soon swept through with models for an inflationary expansion phase at epochs characterized by the GUT energy scale. Other BSM ideas were developed to address the presence of the dark matter (see Chapter 3), which is expected to be a new form of extremely weakly interacting matter. The apparent acceleration of the expansion of the Universe at the most recent epochs has, after exhaustion of the more plausible astrophysical explanations, led to the consideration of a different type of scalar fields leading to forms of dark energy as an explanation for this dynamic bulk behavior.

Then, if we ratchet up the energies to even much larger levels than GUT energies, simple dimensional arguments strongly suggest that quantum effects will become comparable to gravitational ones. This occurs at the Planck energy scale, $E \sim (\hbar c^5/G)^{1/2} = 1.2 \times 10^{19}$ GeV, where one might expect all four of the known forces to become unified. This leads to the need to formulate a (so far unfinished) quantum theory of gravity. String theory is the most widely considered approach towards achieving this goal, while quantum loop gravity is a different approach which is also being considered (e.g. [5–7]). These theories address what happens at the earliest conceivable instants in the Universe, as well as what happens inside black holes near the classical central singularity, which in a quantum theory of gravity is expected to be avoided due to the Uncertainty Principle which introduces an unavoidable fuzziness over energies and times of order $\Delta E.\Delta t \sim \hbar$.

2.4   Into the soup

All of the previous ingredients, quarks, leptons, bosons, atoms, etc., go into making our Universe, as so many ingredients of a Cosmic Soup. The current mainstream scenario is that initially, at times extremely close to the initial instant of the Big Bang, the Universe would have been extremely hot, with temperatures of the order of the Planck energy scale, and it would have been permeated with chaotic space-time fluctuations of the quantum vacuum. These might have already coexisted with, or later transitioned into, quantum fields containing the seeds of what later would become the separate strong, weak, electromagnetic and gravitational fields, the super-symmetric quanta of which inter-converted between fermionic and bosonic states. The gravitational fields would have decoupled very soon after this instant from the rest of the fields. By the time the Universe cooled to temperatures comparable to the GUT energy scale, the strong force would have in turn decoupled from the rest, and by the time temperatures comparable to the electroweak energy scale were reached, the weak and the electromagnetic forces would have decoupled from each other. The Universe would still have been made up of a quark, gluon, lepton and boson soup, which only when QCD-scale temperatures (GeV and above) were reached would have jelled into the recognizable baryons, protons and neutrons that we recognize today. This journey and its aftermath, from the Planck Era to today, is discussed in the next chapter.

3

Cosmology

3.1   The dynamics of the Universe

The present-day Universe appears to be expanding in all directions, as shown by the fact that all distant galaxies and clusters of galaxies appear to be receding from us. This was the first and most obvious piece of evidence indicating that our Universe was initially much denser, leading to the hypothesis of an origin in an initial "Big Bang".

The recession velocities of the galaxies are measured by analyzing the light they emit, which in a spectrograph is seen to contain not only a continuum of frequencies but also discrete frequencies, due to electronic transitions between energy levels of atoms in these galaxies. Such lines have a well-determined laboratory frequency, and when we observe such well-known atomic lines but we see that their frequency is lower (or their wavelength is longer, since wavelength equals speed of light divided by frequency), we infer that the atoms and the galaxy are moving away from us. This effect is called the Doppler shift. A simple everyday acoustic analogy of this Doppler shift is provided by the pitch of an ambulance's siren, which gets lower as the ambulance speeds away from us: the motion away from us "stretches" out the wavelength.

The expansion velocities increase with the distance away from us at a rate which is proportional to the distance, as long as the galaxies are not too far away. This is the famous Hubble law, written as

$$v = H_0.D \tag{3.1}$$

where v is the recession velocity ($\mathrm{km\,s^{-1}}$; the expression is valid only if v is significantly below the speed of light), D is the distance in megaparsec (see below for the definition of parsec) and $H_0 \simeq 70\,\mathrm{km\,s^{-1}\,Mpc^{-1}}$ is Hubble's "constant", a quantity which is derived from measurements. The expansion appears to be uniform, that is, the velocity at the same distance is the same in any direction that one looks. In addition, the Universe on average appears to exhibit the same degree of matter density distribution in every direction. That is, the Universe is homogeneous and isotropically expanding, and the dynamics of the expansion is understood in terms of the general relativistic theory of gravitation. The conclusion is that at earlier times the distant clusters of galaxies must have been closer to us, and the average density of the Universe must have been higher. If we naively extrapolate the expansion backwards in time, we would reach the conclusion that at some instant around 14 billion years before the present time the Universe should have had an essentially infinite density. This instant of time at which the expansion started is called the *Big Bang* [15], and the inferred present "age" of the Universe, $t_0 \simeq 14$ billion years, is loosely called the Hubble time, since $t_0 \sim t_H = 1/H_0$.

The fact that the Universe as we know it has existed only for 14 billion years[1] means that the light from distant galaxies, traveling at the speed of light $c = 3 \times 10^{10}\,\mathrm{cm\,s^{-1}}$, can only have reached us from a distance of at most 14 billion light years. A light year is the distance traveled by light in a year: 1 light year = (speed of light) $\times$ (1 year) $= 3 \times 10^{10}\,\mathrm{cm\,s^{-1}} \times 3 \times 10^7\,\mathrm{s} \simeq 9 \times 10^{17}$ cm. Astronomers use for historical reasons a slightly larger unit, the *parsec*, abbreviated pc: the parsec is $1\,\mathrm{pc} = 3 \times 10^{18}$ cm $\simeq 3$ light years. Thus, our visual "horizon", also called the Hubble horizon, is $D_H \simeq$ (speed of light) $\times$ (Hubble time) $\simeq 3 \times 10^{10} \times 1.4 \times 10^{10}\,\mathrm{yr} \simeq 4.2\,\mathrm{Gpc}$ (where Gpc stands for gigaparsec $= 10^9\mathrm{pc}$, that is a billion parsec).

Because of the finite speed of light, we don't observe anything beyond the Hubble horizon, so in principle we cannot probe what is the extent of the Universe. Is it infinite, or is there an "edge" to the Universe, and if so, what is beyond? And even if it is infinite, what lies beyond our Hubble horizon? These are difficult questions. However, from the fact that the Universe is observed to be isotropic and homogeneous as far as we can reach with our instruments, there is one fairly safe conclusion we can reach. This is that if somehow we were able to travel to any point in the periphery of our currently observable horizon, or if we were able to see with the eyes of an observer currently located at such a point which is in *our* periphery, such an observer would see us as if we were at his periphery. But the Universe where we are would not look any different from the Universe in any other direction in which this observer would

---

[1] This is also verified by the geological record, nuclear isotope chronology and various other methods.

look, including directly away from us. That is, at another Hubble distance D_H measure from this observer, that is two Hubble distances from us, the Universe should look the same again. This process can be repeated *ad infinitum* in all directions, and one concludes that there is no reason to infer the existence of a "center" of the Universe, or for that matter, an "outer edge" of the Universe. The Universe should look the same, no matter how far away from us, at any instant of time.

One can visualize this by means of a simple analogy. We live in three spatial dimensions (3-D), but imagine that we lived in a two-dimensional (2-D) flat plane, a gigantic pampa. Suppose this plane is being stretched in all directions uniformly, so that observers in this plane move away from each other at velocities proportional to their increasing distances. Since light travels at a finite speed, at time t since the start of the expansion each observer would only know about those other observers who are inside a distance ct from them. This distance increases with time, but nevertheless at each time one sees only what is within a finite circle of view. However, from God's point of view, or from the point of view of a 3-D observer outside this 2-D world, one would see no difference between the expansion behavior of any point compared to that of any other point in the 2-D plane, even if the 2-D plane is infinite in extent. It has no center, and no edge that one could speak of.

There is, however, another possibility for the Universe to be finite, even though there is no edge to it. Consider, instead of the analogy of a 2-D flat plane, a 2-D spherical world. That is, the observers live on two dimensions, but these two dimensions are on the surface of a sphere, like bugs on the surface of the Earth or on the surface of a ball, who are unaware of the existence of up and down, only of the two dimensions of the ball's surface. If the ball starts inflating, all distances between various bugs on the ball will increase, at a velocity proportional to the distance (as in the Hubble expansion), but the ball has a finite extent. In principle, a bug or a ray of light forced to follow the surface of the ball could go all the way around the ball, and come back to its point of departure, without having experienced any "edge", nor for that matter having experienced any departure from homogeneity and isotropy: every point on the ball looks the same as every other point.

In fact, General Relativity naturally allows for the possibility of a 3-D Universe which is either infinite, or is finite but has no center and no edge. Before going into this it is useful to consider a simpler Newtonian version of this line of thought which illustrates more intuitively how this could come about. One observes galaxies, that is masses, and one observes their speeds of recession. If one knows all the mass inside the Hubble horizon, one knows in principle the amount of gravitational potential energy (the energy in gravitational attraction of all the masses on each other). Knowing the individual velocities and masses of the galaxies, one also knows the total kinetic energy of motion of these galaxies. They are moving outwards now, but if the total gravitational potential energy exceeds the kinetic energy, we can infer that the galaxies will have to slow down their expansion, and at some point they should turn around and start re-contracting. This is the same kind of thing which happens with a rock thrown up in the air: even though initially it moves up, it will slow down and eventually it falls back. However, if one threw it fast enough, it could escape the Earth and fly "forever". In the same way, a rocket equipped with a strong enough booster can achieve a velocity large enough to escape the gravitational attraction of the Earth, and never come back. If the gravitational energy just equals the kinetic energy, the rocket can continue moving away forever. In the case of the Universe, if the initial kinetic energy of expansion exceeds or is just equal to the gravitational energy, the Universe can expand forever, and as time passes the Hubble horizon $D_H \sim ct$ goes to infinity. However, if the initial kinetic energy is less than the gravitational energy, as time grows the expansion slows down and turns to collapse, and there is only a finite time available for observations before the "big crunch": the Hubble horizon does not grow to an infinite extent but only to a maximum extent, after which it shrinks, and the Universe available to our observations is finite.

In General Relativity, as discussed in Chapter 2, gravity is described as a distortion of space-time caused by the masses, such that all massive bodies or particles naturally move in it following the curvature of space-time, just like children glide along the curved path of a playground slide. The total amount of mass in the Universe is what determines the curvature of space-time. In Einstein's equations for a homogeneous isotropically expanding Universe, also called a Friedman model of the Universe, there is a critical mass density (actually mass-energy density) which is $\rho_c = (3H_0^2/8\pi G)$, where G is Newton's constant of gravity, and H_0 is the current value of the Hubble constant. This critical density is

$$\rho_c \simeq 10^{-29} \, \text{g cm}^{-3} \tag{3.2}$$

which is incredibly small.[2] This density is the equivalent of the mass of 10 protons per cubic meter, or 10 mg – which is 1/50th of the mass of an aspirin! –

---

[2] The exact value is $\rho_c = 1.88 \times 10^{-29} h^2 \, \text{g cm}^{-3}$, where the Hubble constant, whose value has been debated over the years by astronomers, is normalized to a value h defined as $h = (H_0/100 \, \text{km s}^{-1} \, \text{Mpc}^{-1}) \simeq 0.7$; the current "best" measured value is $H_0 \simeq 70 \, \text{km s}^{-1} \, \text{Mpc}^{-1}$, so $h \simeq 0.7$

in a volume comparable to that of the Earth, except that only 1/7 of this mass is actually in protons while 6/7 is dark matter. Thus, if you look up at the night sky and feel a sense of emptiness, well, you are fully justified. Although it is non-zero, this density is so minute that the emptiest "vacuum" achievable with the best pumps in the laboratory would appear to be very dense by comparison.

Of course, this density is just the mean amount of mass per unit volume. This does not mean that there is very little mass in the Universe. There is in fact quite a lot, since there are very many "volumes of the Earth" in the Universe, and consequently very very many milligrams in it. Just inside our visible Universe, that is inside a light horizon of 4.2 Gpc, there are about 80 billion galaxies, and on average each galaxy has 400 billion stars of roughly one solar mass each, so within the visible Universe there are (billions and billions does not cover the concept) roughly 5×10^{22} stars – 50 sextillion stars, if you wish – amounting to a total of 3×10^{79} protons, or roughly 5×10^{52} kg inside the whole visible Universe. Since normal baryonic matter is at most 4% of the total Universe, adding the mass equivalent of the dark matter and the dark energy (see below) we come to a total mass within the visible Universe of 10^{54} kg (that is, 1 followed by 54 zeros kilograms). Now, if you were to grind up all of this mass very fine, finer than for espresso, and you spread it out evenly over the visible Universe, you would get an average mass per unit volume comparable to the critical density ρ_c of eq. (3.2). This density is so low because even though there is a lot of mass, the visible Universe has a huge volume – about 10^{80} (1 followed by 80 zeros) cubic meters.

The reason ρ_c is called the "critical" density is that the geometry of the Universe we live in depends on whether the actual average mass-energy density of the Universe is above, equal to or below this density ρ_c. This is conventionally expressed through a cosmological density parameter Ω_0, which is the ratio of the actual mass-energy density to the critical one,

$$\Omega_0 = \frac{\rho_0}{\rho_c}. \tag{3.3}$$

If the density of the Universe is precisely equal to the critical density, $\Omega_0 = 1$, this implies that the curvature of space-time is "flat", like the familiar Euclidean space of geometry textbooks, i.e., it is not curved at all (see Fig. 3.1, bottom panel) [16]. In such a geometry the sum of the internal angles of a triangle is the usual 180°, and two parallel lines extended to infinity always remain parallel. Such a Universe extends infinitely in all directions, and the distance between any two observers (or clusters of galaxies) will continue to grow in time forever. Rescaling the current Hubble radius $D_H = c/H_0$ through an arbitrary scale factor R which is taken to be unity now, this scale radius will grow forever,

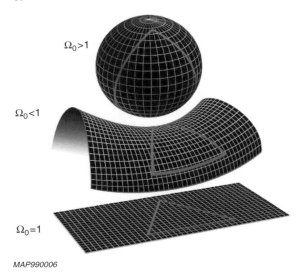

$\Omega_0 > 1$

$\Omega_0 < 1$

$\Omega_0 = 1$

MAP990006

Figure 3.1   Three possible types of Universe geometry: a closed $\Omega_0 > 1$, an open $\Omega_0 < 1$ and a flat $\Omega_0 = 1$ model.
Source: NASA.

at a rate which approaches a constant velocity of expansion, as shown by the curve labeled $\Omega_M = 1$ in Fig. 3.2.

However, if the density of the University is larger than the critical value ρ_c, $\Omega_0 > 1$, the curvature of space-time is positive, and the Universe closes in upon itself, like the 2-D beach ball analogy (Fig. 3.1). This is a finite Universe, in which the sum of the angles of a triangle exceeds 180°, and the distances between observers initially grow in time, but eventually start to decrease towards zero. In such a Universe the expansion eventually turns around and it recollapses, as shown by the curve for $\Omega_M = 6$ of Fig. 3.2.

On the other hand, if the density of the Universe is less than the critical density, $\Omega_0 < 1$, the curvature of space-time is negative, which in the 2-D analogy corresponds to the surface of a saddle (see Fig. 3.1). In this case the Universe is also infinite, but the interior angles of a triangle are less than 180°, while the distances between observers grow faster in time than in the case of the flat Universe. In this case, as in the flat $\Omega_0 = 1$ case, the scale factor R grows forever, but faster than in the flat case. This is shown by the curve for $\Omega_M = 0.3$ in Fig. 3.2.

It is conventional to define the scale factor to be $R = 1$ at the present time, and for a given Ω there is a unique relation between the age of the Universe t and the scale factor R (Fig. 3.2). Since light travels at a finite speed c, an object at a distance D is observed as it was when the Universe had an age of $t = D/c$ (ignoring relativistic corrections to the definition of distance, which are small

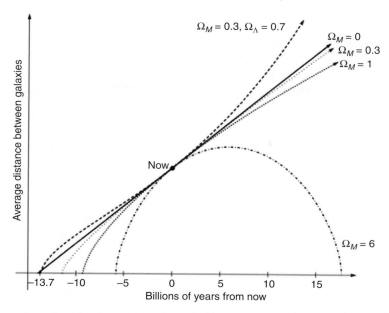

Figure 3.2 Schematic behavior of the cosmological scale factor R (y-axis) versus time (x-axis), taking here $t = 0$ as the present time, for different amounts of dark matter Ω_M and dark energy Ω_Λ. The five model Universes have the same slope (Hubble velocity) at the present epoch $t = 0$, $R = 1$. Shown are a closed model $\Omega_M = 6$ which recollapses; a critical (flat) $\Omega_M = 1$ model without vacuum energy which expands forever at a decelerating rate that approaches coasting; two sub-critical open models, one with $\Omega_M = 0.3$ and one empty $\Omega_M = 0$, both without vacuum energy, which expand forever at a rate that approaches coasting; and a critical (flat) $\Omega_T = \Omega_M + \Omega_\Lambda = 1$ model with vacuum energy, which turns to an accelerated expansion, sloping upwards.
Source: SNAP team website.

when the latter is not too close to the Hubble radius). The age of the Universe t is related to R, and at the same time it provides a measure of the distance to an object from which light took a time t to reach us. For this reason, it is conventional to define a quantity called the redshift z, which depends on the radius R at a particular time t,

$$z = \frac{1}{R} - 1. \tag{3.4}$$

That is, z measures how much the radius of the Universe R at a given time t differs from the present radius $R = 1$ at the present time $t = t_0 \sim t_H$. At the present time, $R = 1$ and $z = 0$, while at an earlier epoch when the Universe was half its present radius, $R = 1/2$ and the redshift is $z = 1$. Thus, the redshift serves both as a label for the age of the Universe t when it had a particular radius R, and as a measure of the distance D of an object from which light started to travel towards us when the Universe had an age of t. Higher redshifts z imply more distant objects, and earlier times in the history of the Universe.

A determination of the total amount of mass in or around the galaxies is difficult, because besides the normal "baryonic" matter in galaxies (protons, nuclei and their electrons, some of which is visible through the electromagnetic radiation it emits) there is also the dark matter component, which emits no radiation known so far, but whose presence is inferred only from its gravitational effects. The amount of dark matter exceeds the normal baryonic matter significantly: the normal baryonic matter is $\sim 4\%$ while the dark matter is $\sim 22\%$ of the critical density, so the total amount of "matter", baryonic plus dark, is about 26% of the critical density ρ_c (e.g. [17]). However, this does not yet mean that the Universe has negative curvature, because it is the mass-energy density that has to be compared to the critical one.

And, as we now know, there is in addition to the baryonic and dark matter another even larger amount of energy in the Universe, the "dark energy", of unknown origin, which is causing the distant clusters of galaxies to expand at an increasing rate. The amount of dark energy plus that of dark matter and baryonic matter just about equals the critical density, making the Universe flat, as indicated by current observations. The dark energy has an effect similar to that of introducing a cosmological constant into the general relativistic equations describing the evolution of the Universe. This dark energy, or vacuum energy as it is also called, appears to have become dominant only in the recent cosmological past, at redshifts $z \lesssim 0.5$ [18]. The expansion rate of the Universe is somewhat modified by dark energy for times close to the recent past, and will be strongly modified in the future, as shown in Fig. 3.2 for the curve $\Omega_M = 0.3, \Omega_\Lambda = 0.7$. A Universe with these parameters, which are suggested by current observations [17], has a flat geometry since $\Omega_T = \Omega_M + \Omega_\Lambda = 1$, but the presence of dark energy, $\Omega_\Lambda \simeq 0.7$, causes it to expand at an accelerated rate, starting at epochs close to the present.

3.2   The primordial fireball: a particle cauldron

Since the Universe is observed to be expanding, at earlier times (higher z) it must have been denser. If we imagine regressing backwards in time, this is the equivalent of compressing the Universe, and we know from everyday experience that compressing a gas, or anything for that matter, leads to heating it up. The simplest example is a bicycle pump, in which the air gets hotter as we compress it, as we verify by putting our hand on the side of the pump.

Reversing the process, as a gas expands, it cools down (again, as we release the bicycle pump handle, the air expands and the pump wall feels cooler). Thus, in its earlier stages the Universe was denser and hotter, the more so the further back we go in time.

Currently most of the "normal" detectable matter in the Universe consists of hydrogen (90% by number), helium (10% by number), and traces of heavier elements such as carbon, nitrogen and the rest.[3] For most of the recent past, these elements were in their neutral form, that is, with electrons attached to the corresponding nuclei, in the form of electrically neutral atoms.

The other ubiquitous constituent of the present Universe is photons. Photons, such as starlight, pervade the Universe, as do other types of photons, such as X-rays and gamma-rays from some types of galaxies, infrared photons from new stars in young galaxies, radio photons from some types of galaxies, etc. However, the most abundant type of photons in the Universe have a wavelength in the millimeter range, between the radio and the far infrared wavelengths: these are the so-called *cosmic microwave background* (CMB) photons. Their density exceeds that of any other type of photons: there are around 430 of them per cubic centimeter. Their frequency spectrum is thermal, which means that it is a continuous spectrum such as would be emitted by a black body heated to a certain temperature. The average wavelength of a CMB photon is roughly a millimeter (0.1 cm), and the black-body temperature corresponding to the CMB photons is $T = 2.73\,\text{K}$ in the Kelvin absolute degree scale. That is 2.73 degrees above the absolute zero temperature, below which no further cooling is possible.[4]

Photons are particles, whose collection can be thought of as a gas. They are contained in the Universe, and have nowhere else to go, so if the average density of the Universe was higher earlier on, these photons would have been compressed into a smaller volume, and their equivalent temperature must have been higher, like any other compressed gas. Equivalently, if their wavelength is now 0.1 cm, earlier on these waves must have been squeezed into a smaller volume, which means that their wavelength must have been shorter, since we must still have the same number of wave peaks and wave troughs that we had before compression. Using the second law of thermodynamics, one can show that the equivalent temperature of a photon gas increases inversely with the

---

[3] Of course, planets and our own bodies consist mainly of heavier trace elements such as carbon, oxygen and others, but these are minute fractions of the baryonic mass of the Universe, of which we are very specialized sub-units.

[4] The Kelvin temperature scale at normal pressure assigns to absolute zero temperature the value $0\,\text{K}$ ($\equiv -273°\text{C} \equiv -459.67°\text{F}$), while water freezes at $273\,\text{K}$ ($\equiv 0°\text{C} \equiv 32°\text{F}$) and water boils at $373\,\text{K}$ ($\equiv 100°\text{C} \equiv 212°\text{F}$).

size of the volume in which it is contained. Thus, as the scale R of the Universe varies (e.g., Fig. 3.2), the CMB temperature varies as $T \propto 1/R$, and T was higher earlier on. For the same reason, the mean CMB photon wavelength varies as $\lambda \propto R$, and the mean frequency ν of a CMB photon (which is related to the wavelength via $\nu = c/\lambda$, where c is the speed of light) increases as we go back in time proportionally to the temperature, $\nu \propto T \propto 1/R$. The energy of a photon is proportional to its frequency, which can be understood if we think of the photon as a packet of oscillating electric and magnetic fields – the faster they oscillate up and down, the more energy is involved; just think of yourself whipping a cord up and down. The proportionality between the photon energy ε and the frequency ν is $\varepsilon = h\nu$, where $h = 2\pi\hbar = 6.625 \times 10^{-27}$ erg s is Planck's constant. Thus, the photon energy also increases as $\varepsilon \propto 1/R$, as we go back in time. The modest millimeter photons of today's CMB must have earlier been optical photons, and even earlier UV photons, X-ray photons, and so on.

The matter, such as hydrogen, heavier elements, dark matter, etc. must also have been hotter earlier on, just as if it had been inside a container whose volume was shrinking. This means that the collisions among baryonic matter and photons must have been more frequent, since they had less distance to travel to encounter each other. The photons were also more energetic earlier on, and if we consider an epoch when the scale of the Universe was about 1000 times smaller than at present, $R \sim 10^{-3}$, the energy of the photons would have corresponded to a temperature of $T \sim 2700$ K. Photons of such energy are capable of stripping electrons out of neutral atoms, a process called ionization. An atom stripped of one or more of its electrons is called an ion. The hydrogen atom has only one electron, so at earlier times or smaller R corresponding to redshifts $z \gtrsim 10^3$, hydrogen would be fully ionized. Similarly, other elements such as helium, carbon, etc., where the electrons are more tightly bound, become ionized at even higher temperatures or higher redshifts. In the fully ionized Universe at $z \gtrsim 10^3$, the photons scatter off the free electrons, which are now plentiful. The protons, i.e., the ionized hydrogen nuclei, do not scatter as effectively as the electrons, but on average the electron and proton distributions follow each other due to preservation of overall electric charge neutrality, so the radiation (photons) and the matter are tightly coupled at early times corresponding to $z \gtrsim 10^3$. In the normal forward progress of time, they must become decoupled at times corresponding to $z \lesssim 10^3$ – this is when the free electrons disappear, by becoming attached to protons to make neutral hydrogen atoms, which do not scatter photons efficiently. Thus, an important watershed in the history of the Universe occurs at redshifts $z_{dec} \simeq 1370$, which is called the decoupling epoch, or essentially the electron and ion recombination epoch. It corresponds to an age of the Universe of $t_{dec} \sim 1.8 \times 10^5$ yr $\simeq 5 \times 10^{12}$ s (Fig. 3.3).

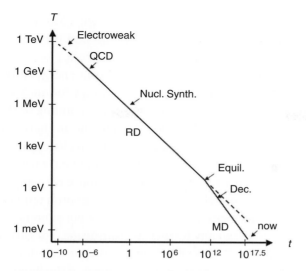

Figure 3.3 Thermal history of the Universe, showing temperature T (in units of energy per particle) versus time t (in seconds). The radiation and matter-dominated eras are indicated as RD and MD, showing also the electroweak transition, the QCD, the nucleosynthesis, the matter–radiation equilibrium and the radiation decoupling epochs, as well as the present one.

Continuing the backward journey in time, as long as particles are neither created nor destroyed, the number density and the rest-mass density of baryonic and dark matter particles increases inversely with the volume, which is $\propto 1/R^3$, and hence the rest-mass density depends on redshift as $\rho_m \propto (1+z)^3$. This increase with redshift is characteristic of non-relativistic particles, which applies to baryons and dark matter for not too early epochs. Photons, however, are relativistic particles at all epochs, since they always travel at the speed of light. They have no rest-mass, but they have an energy per particle $\varepsilon = h\nu = hc/\lambda$, and since wavelengths are compressed as we go back in time (up in redshift) because the size of the Universe was smaller, the photon energy per particle goes up with redshift, $\varepsilon \propto (1+z)$. In the absence of creation and destruction of photons, their number density also goes up as $n_\gamma \propto 1/V \propto (1+z)^3$. Thus the photon energy density goes up as $\rho_\gamma = n_\gamma.\varepsilon \propto (1+z)^4$, one factor of $(1+z)$ faster than the energy density of baryonic plus dark matter (or matter, for short). At present the ratio of photon to matter energy density is approximately $\Omega_\gamma/\Omega_m \simeq 1.7 \times 10^{-4}$, so the energy densities of matter and photons become equal at $z_{eq} \simeq 1.35 \times 10^4$, when the Universe was about $t_{eq} \sim 4.4 \times 10^3$ yr $\simeq 1.3 \times 10^{11}$ s old (Fig. 3.3).

At epochs earlier than $z_{eq} \sim 1.3 \times 10^4$, the energy density of the Universe would have been dominated by the radiation, its temperature evolving as

$T_\gamma \equiv T = T_{\gamma,0}(1 + z)$, where $T_{\gamma,0} = 2.73\,\mathrm{K}$ is the present CMB temperature. At a redshift $z \sim 10^9$, the temperature T of the photons and baryons is about 10 billion Kelvin, $T \sim 10^{10}\,\mathrm{K}$, when the Universe was about $t \sim 1$ s old, i.e., one second after the Big Bang. This temperature or mean energy of the photons and baryons is about 1 MeV, which is the typical energy for nuclear fusion reactions. Nuclei colliding at the thermal velocities corresponding to this temperature can undergo fusion, as they do inside the Sun and in thermonuclear bombs or fusion reactors, and can create more complex nuclei out of simpler ones. At even earlier epochs the photon temperatures would have been so high that when they collided with any fused nuclei they would have rapidly dissociated them, but after t approximately a few seconds the photon temperatures become mild enough to spare them. After $t \sim 100\,\mathrm{s}$ the encounters between nuclei become too rare to undergo any further nuclear fusion. For this reason, the epoch of $t_{BBN} \sim 1 - 100\,\mathrm{s}$ is called the Big Bang nucleosynthesis (BBN) epoch, because it is thought that it is during this epoch that most of the lighter nuclei found in the Universe, such as helium and lithium, were synthesized via fusion of single nucleons like protons and neutrons.

So what happened even earlier? Admittedly $t \sim 1$ s is already extremely early, but time can be sub-divided even more finely, and temperatures higher than 1 MeV can be expected. For example, at $t \lesssim 10^{-5}\,\mathrm{s}$ the temperature would have been higher than $T \sim 1\,\mathrm{GeV}$, which means that collisions between nucleons, or nucleons and photons, would have led to the production of sub-nuclear particles such as pions, kaons, mesons, etc., as in laboratory accelerators. This is the QCD era (Fig. 3.3), where quarks and gluons start to become major players. At even earlier times one would have had temperatures $T \sim 100\,\mathrm{GeV}$, corresponding to particle kinetic energies comparable to the weak W and Z boson masses. This occurs at an epoch $t_{EW} \sim 10^{-10}\,\mathrm{s}$, called the electroweak epoch, since above this temperature or energy the weak and electromagnetic interactions become unified in a single electroweak theory. Even further back in time, at $t < t_{LHC} \sim 10^{-14}\,\mathrm{s}$, the temperatures would result in collisions whose energy exceed the maximum design energy of $E \sim 14\,\mathrm{TeV}$ of the LHC in CERN. This is deep into the world of the quarks, gluons, fermions, bosons and leptons, etc. discussed in the previous chapter. This is the reason why particle physicists study the cosmology of the early Universe, and why astrophysicists delve into particle physics [1, 19].

3.3   Into the unknown: the GUT and Planck eras

As we regress further into increasingly smaller times the story gets more exotic, and heated. At times much earlier than the LHC epoch t_{LHC} mentioned above, $t \sim 10^{-36}$ s, one would eventually encounter temperatures of order $T \sim T_{GUT} \sim 10^{15}-10^{16}$ GeV, where it is expected that the strong interactions become unified with the (already unified) electroweak interactions in a single GUT. This leaves only the gravitational interaction as the last remaining independent interaction, which will require even higher temperatures to become unified. There are at present several different GUTs, which cannot be tested directly because the GUT energy far exceeds the nominal range of accelerators such as the LHC. It is hoped, nonetheless, that the LHC may be able to provide at least some indirect pointers towards a correct GUT, for instance by finding lower energy particles which decayed from more energetic particles associated with the GUT scale. It is also possible that the ubiquitous dark matter (see Section 3.4) has its origins in the GUT era, and either direct or indirect low energy detection of dark matter signatures might also lead to constraints on hypothetical GUT.

The most difficult and fascinating problem is what happens at epochs earlier than the GUT epoch, when the Universe was so dense that the typical temperatures would have values corresponding to the Planck energy $E_{Pl} = (\hbar c^5/G)^{1/2} \simeq 1.2 \times 10^{19}$ GeV. The corresponding Planck epoch is $t = (\hbar G/c^5)^{1/2} = 5.4 \times 10^{-44}$ s. At these energies quantum and gravitational effects become comparable, and it is at these energies that we expect that gravity might become unified with the electromagnetic, weak and strong forces in an ultimate Theory of Everything (TOE). Leading candidate TOE at present include string theory and quantum loop gravity theory, in both of which quantum and gravitational effects are comparably important [7]. The Uncertainty Principle says that $\Delta E \Delta t \gtrsim \hbar$, so in fact one might expect that the classical singularity of General Relativity at the origin of the Big Bang is avoided, due to the quantum fuzziness, being replaced instead by quantum fluctuations. The Universe might in fact have experienced a bounce at this epoch, from a previous epoch of contraction leading to the current epoch of expansion. This might, if we have just a tiny excess of energy above the critical value, result in an oscillating Universe with an endless series of bounces. Physics at this epoch is, however, still a very preliminary and tentative groping towards results.

3.4   Inflation, dark energy and dark matter

One common feature of many GUTs is that they appear to lead to a phenomenon called inflation, which has a huge impact on cosmology. Inflation is a period of extremely rapid expansion, where the scale R of the Universe grows exponentially with time over a brief period of time, roughly around $t \sim 10^{-36}$ s-10^{-34} s. This exponential growth is in contrast to the more moderate power law growth in time of later epochs. The exponential growth causes the length scale R to increase by a factor of about 60 "e-foldings" or $e^{60} \simeq 10^{26}$ in those two brief decades of time, a huge stretching of all length scales. This phenomenon is ascribed to a new type of vacuum scalar field present at this time, which results in a negative pressure. Normal particles generally have a positive pressure, and positive pressures have a positive energy, but a negative pressure can cancel out the positive kinetic energy of the field. Grossly oversimplifying the situation, gravity, which acts on the total mass-energy density of everything present, does not have anything to act upon, due to the cancellation between the pressure and the thermal energy contributions, so in the absence of gravitational attraction an exponential expansion sets in.

Inflation provides cosmology with a natural explanation for why the Universe appears today on average so smooth and homogeneous in all directions, and for why the curvature of space-time appears to be essentially flat. Again grossly oversimplifying the argument, one can see that even if the Universe was initially chaotic, with density bumps and depressions, an episode of inflationary expansion by $\sim 10^{26}$ represents such a huge stretching of all length scales that all bumps and troughs get "ironed out", practically smoothed out to extremely small values compatible with the homogeneity now observed. Similarly, if the Universe initially had positive or negative curvature (equivalent to a 2-D sphere or saddle), a stretching by $\sim 10^{26}$ would cause any region which encompasses our present-day horizon to appear essentially flat. This is because what today is our huge horizon was, at the time inflation started, an almost infinitesimally small region of space, and an infinitesimally small region of any curved surface appears flat. The expansion by the huge factor of 60 e-foldings makes this small flat space-time region expand into an extremely large flat space-time region, as the Universe appears today.

The particle physics of inflation is speculative at the moment, and involves heavy doses of beyond the Standard Model particle physics. The new vacuum field which is thought to give rise to inflation in the period $10^{-36}-10^{-34}$ s is a particular type of scalar field, other variants of which appear in other areas of physics. The inflationary field involves aspects of GUT, which are plausible but unproven, and various variants have been proposed. Interestingly, at redshifts $z \lesssim 0.5$, which is earlier but fairly close to the present time, there appears to be another type of vacuum field. This is invoked to explain the accelerated expansion of the Universe observed at these low redshifts, detected through observations of distant Type Ia supernovae, microwave background observations, etc. This accelerated expansion is described as an additional vacuum energy term, i.e., the previously mentioned "dark" energy (Section 3.1), which is not associated with any particles. The characteristic energy of this field is so many orders of magnitude smaller than the energies inferred in inflation that it must be an altogether different field. However, phenomenologically they both appear to behave like vacuum fields, sharing the similarity of leading to an exponential expansion. The exact nature of both fields is mysterious, although the consequences of dark energy on the dynamics appear clear (or at any rate the accelerated expansion ascribed to it appears clear), while the consequences of an inflationary phase are so attractive as to be compelling.

The other mystery is why dark matter, which also appears to involve BSM physics, provides an energy density which at the present time is so close to the apparently unrelated dark energy density. Dark matter seems to be a pressure-less gas of massive, non-relativistic particles (see Chapter 4), while dark energy appears to be due to the vacuum quantum fluctuations of an unknown scalar field,[5] leading to a negative pressure. Dark energy, as far as we can tell, is completely non-interacting with any known forms of matter, except through its dynamical effects on the Universe's expansion, whereas dark matter at least is expected to be (extremely) weakly interacting.

Dark matter, from its observational properties, must consist of non-relativistic, pressureless, gravitating massive objects or particles. In principle it could be small black holes, low mass stars or planets which have no nuclear reactions (dubbed MACHOs, or massive compact halo objects), or any small aggregates of matter ("bricks") whose size is larger than optical wavelengths so they would have escaped detection. However, such massive macroscopic objects have been searched for, unsuccessfully (e.g., through gravitational micro-lensing and other types of searches) and the only remaining conclusion is that dark matter must consist of some form of new elementary particle. These particles cannot have an electromagnetic charge, otherwise their interactions would have led to their detection, and for the same reason they are not expected to be subject to the strong force. They could however be endowed with interactions whose strength is comparable to that of the standard weak force, even if it is not the same force. There are in fact well-motivated BSM theories, in particular supersymmetric (SUSY) theories (Chapter 12), which predict dark matter particle candidates having such "weak" interactions, generically called WIMPs (weakly interacting massive particles). The experimental search for WIMPs is an intense field of activity, using both direct methods in the laboratory which attempt to detect the impact of WIMPs and indirect methods such as astrophysical measurements of secondary products of WIMP decays or of their interactions with other normal particles. This is discussed further in Chapter 12.

---

[5] A scalar field is a field which depends on only a single quantity at each position in space-time to denote its strength, unlike a vector field which requires three quantities, one for the strength and two angles for its direction.

4

Cosmic structure formation

4.1   The perturbed Universe

As discussed in the previous chapter, our Universe is at present in the midst of a smooth and uniform expansion in all directions, at least if observed at low spatial resolution (i.e., averaged over scales of hundreds of megaparsecs) From general relativistic cosmology, we know that this expansion would appear the same to any observer located anywhere else in the Universe. That is, we are not in any way in some kind of privileged position at the center of the Universe, but rather we are just one of the folks, along for the ride.

This expanding tapestry of the Universe, which appears so smooth on average, is however punctuated by the presence of regions of enhanced mass-energy density, which evolve and become increasingly more compact and complex as time goes on. It is thought that these density inhomogeneities originate from small initial density perturbations, which are regions where the density is slightly above average, $\rho = \bar{\rho} + \Delta\rho$, where $\bar{\rho}$ is the average density and the density excess $\Delta\rho$ is much smaller than the average. Under the action of gravity the small density excesses grow to larger amplitudes, until eventually the perturbations become significant compared to the average density, $\Delta\rho \simeq \bar{\rho}$, described by saying that the perturbations have become "non-linear". At this point the self-gravity of the perturbation becomes stronger than the expansion of the average Universe (since its density, and hence its gravity, is larger than that of the average). This results in the fluctuations "decoupling" from the rest of the background Universe, and evolving on their own as a separate little Universe. If the background Universe is, for instance, a flat $\Omega = 1$ Universe, any perturbation produced by a local overdensity leads to a region represented by an $\Omega > 1$ Universe model, which will continue to expand only for a finite time and eventually turns around, then starts to recollapse upon itself. This leads to gravitationally bound objects, which are the progenitors of the clusters of galaxies, galaxies and stars that we observe today. The details of the sequence of events are, however, complicated and only approximately understood.

The most important aspect of the process is that after the epoch of matter–radiation equilibrium, most of the mass-energy is in the form of dark matter, and even before recombination (i.e., before the electrons recombine onto protons to make neutral atoms), the dark matter dominates the gravitational field of the Universe and also that of any density perturbation in it [20]. Since the dark matter does not "feel" the radiation, regions with an excess of dark matter relative to the background start to slow down, and eventually recollapse (see Fig. 4.1). The dark matter is, from all indications, non-relativistic; that is, it has no pressure. Thus, when dark matter particles recollapse onto themselves, they are not stopped by their own pressure (they are "collisionless") and they go right through each other. They overshoot, and like a pendulum, eventually they turn around and around again. In the process, the gravitational field varies chaotically, and this acts as a damper on the dark particle motions, which come to a quasi-thermal equilibrium in a few dynamical times satisfying the Virial theorem, which states that twice the kinetic energy of the particles equals their gravitational potential energy. This process

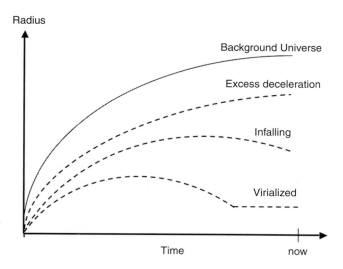

Figure 4.1   Deceleration and collapse of density perturbations relative to the evolution of the background Universe.

is called virialization, the equilibrium outer radius being half the radius at turnaround.

This paradigm, which appears to be supported by current observations, is called the Cold Dark Matter scenario (CDM, for short). Including the effects of a vacuum energy or a cosmological constant Λ alters somewhat the dynamics in the late stages, for redshifts $z \lesssim 0.5$, and this is referred to as the LCDM scenario, where the L stands for Lambda (Λ). The process involves a hierarchy of semi-chaotic sub-structures cascading through a continuum of mass scales, arranging themselves under the action of gravity into ever larger condensations. It also involves, on the other hand, the formation of much smaller, extremely concentrated islands of mass and energy, located inside the larger scale structures, like dense raisins in a pudding. The smaller, non-relativistic dark matter structures collapse and assemble first, within the increasingly larger scale "parent" structures encompassing them. This is because the larger the structure the smaller is the initial average perturbation amplitude (or density contrast), so the larger perturbations collapse after the smaller ones contained in them, and then the even larger ones, and so on. These dark matter dominated structures are called dark matter halos, for reasons explained below.

4.2   Large scale structure formation

As the various mass scales collapse, including those comparable to proto-galaxies (total masses $10^9-10^{12}M_\odot$), the baryons represent a minor constituent which falls into the potential wells provided by the associated dark matter. The dark matter, being collisionless, passes through itself and re-expands until it reaches its approximate virial equilibrium configuration, which is typically spherical if there is not much angular momentum, or oblate spheroidal if the rotation is substantial. These have the shape of an extended dark matter halo, resembling the stellar halo of our and other galaxies, but more extended. The dark matter halos which turned around (collapsed) at a redshift z_c can be shown to end up with a virialized (equilibrium) density $\rho_c \simeq 200\rho_0(1+z_c)^3$, where ρ_0 is the Universe's average mass density today. That is, the ones that collapsed earlier (at larger z) are denser than the ones that collapsed later.

Within the LCDM scenario outlined above, large scale structures in the Universe such as galaxies and clusters of galaxies form hierarchically. Small dark matter (DM) halos collapse first, within whose potential well baryonic gas gives rise to stars resulting in small galaxies. Such small DM halos tend to occur first where there is an underlying larger scale overall density enhancement which gives it a little extra boost, leading to a "biasing" in the formation process. In the course of time the smaller DM halos are expected to merge, before or after stars formed in them, giving rise to larger DM halos and larger galaxies. Even later, the overarching larger density enhancements would themselve become "non-linear", giving rise to a distinct cluster of galaxies. Some of the galaxies and their DM halos will collide and merge, giving rise to a few massive central galaxies in the cluster, while other smaller galaxies would gobble up even smaller neighbors, and grow to become "adult" middle class galaxies within their cluster.

In this picture, the large scale regions of high initial overdensity eventually give rise to rich clusters of galaxies, containing hundreds to thousands of galaxies, with some very massive central galaxies. Examples are the Virgo and Coma clusters of galaxies. A numerical simulation showing such large scale structures is shown in Fig. 4.2. In other, smaller regions, fewer galaxies are expected to form, leading to less compact, small clusters or groups of galaxies. An example is our own "Local Group" of galaxies, which contains a few middle class galaxies (the Milky Way and Andromeda), with an extended group of several

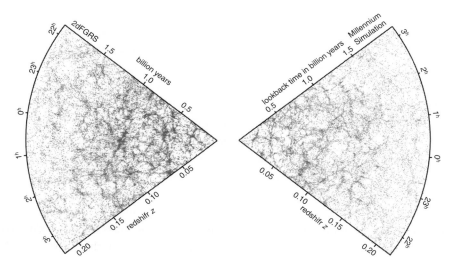

Figure 4.2 Left: observed large scale cosmic structures (i.e., light distribution). Right: numerical simulation of large scale cosmological structures (dark matter) from the Millennium Run consortium. The slices show the distributions in angle and distance (radially, in light years). Note the elongated sheets and knots at the intersection of various collapsed perturbations, the distribution of light being more concentrated than the dark matter [21].

dozen smaller satellite galaxies, which avoided being swallowed up by the larger ones.

What happens during all of this with the baryons? The baryonic gas is a smaller fraction of the total mass than the dark matter, and its dynamics is dominated by the gravitational field of the DM. Thus, the baryons initially follow the DM during the expansion, the turnaround and the early phases of the collapse. As the collapse proceeds, the volume occupied by the DM and the gas decreases and both are adiabatically heated. However, unlike the DM, the baryonic gas is collisional (i.e., its atoms have a significant "cross-section" for interacting with each other as the gas density increases in the collapse), and this gives rise to a further heating caused by collisions between blobs of baryonic gas, leading to shocks which convert the infall kinetic energy into random thermal motion energy of the gas particles. These thermal gas motions lead to collisions between individual atoms and molecules, which excite their electrons to higher quantum energy levels followed by radiative de-excitation; that is, the emission of photons. Thus, within a certain range of densities and temperatures, the baryonic gas can cool efficiently, the photons carrying away much of the acquired thermal energy of the gas, which settles down into a more compact configuration than the parent dark matter halo in which it nests.

Some time after the first galaxies started to form, their larger parent structures started to collapse as well, forming clusters of galaxies. The already formed or forming galaxies in them then virialized against each other. In this cluster collapse process some of the galaxies can approach very close to each other, and more rarely even go through each other, resulting in galaxy mergers which produce a more massive galaxy. In these encounters the dark matter (being collisionless) and the stars (being small) belonging to the individual galaxies pass right through or past their counterparts in the other galaxy, and after a few oscillations they again virialize into a combined, larger DM halo and galaxy. In this manner, some galaxies can grow to very large sizes. Whatever loose gas is present in the merging galaxies undergoes collisions and shocks, settling down into a new gaseous configuration, and a new round of star formation ensues. These are known as "star-forming" galaxies, usually much brighter in the UV and blue light of newly formed massive stars.

The approximate correctness of this picture is given support by several quantitative considerations which follow from it. One of these is an estimate of the expected mass of the luminous galaxies, which are made of stars (i.e., baryons). Clearly, in order to form, structures such as galaxies must have a collapse time which is shorter than the age of the Universe at that epoch. Furthermore, in order for the gas to condense to large enough densities to make stars, the gas must be able to cool in a time shorter than the collapse time. This can be shown to lead to a maximum mass of gas of the order of $M_{g,max} \sim 10^{12} M_\odot$ in gas (adding a DM halo component, this is increased to about $10^{13} M_\odot$ total). This is comparable to the mass of the largest galaxies observed. Larger masses would not have had time to cool in the age of the Universe, and would not yet have condensed into stars [22, 23]. On the other hand, for galaxies whose total mass is too small, the gravitational attraction of the DM may not be sufficient to retain the shock-heated gas, which would evaporate and disperse into intergalactic space. This limit occurs around $10^8 - 10^9 M_\odot$, which is indeed the lower range of observed galaxy masses.

Another test of the general correctness of the LCDM scenario is provided by the phenomenon of gravitational lensing. In General Relativity, all forms of energy (including electromagnetic waves, i.e., light) have an equivalent mass, which is acted upon by the gravity of other masses. Thus, the light rays coming to us from a distant object along a path which passes near some foreground massive object of mass M are expected to be deflected slightly by the latter, by an angle

$$\alpha = \frac{4GM}{c^2 b} \equiv \frac{2R_S}{b},$$
(4.1)

where b is the "impact parameter" (the minimum distance of approach of the light ray to the mass M) and $R_S = 2GM/c^2$ is the Schwarzschild radius defined in eq. (4.2).

This effect, predicted by Einstein in 1916, was verified by Eddington in 1919 for the case of the light from a star grazing the disk of the Sun, where the expected and observed deflection is $\simeq 1.75''$. Now, if the light rays from a distant galaxy or a quasar on the way to us pass through a foreground cluster of galaxies, these light rays will miss most of the luminous cluster galaxies, which are relatively compact, but they will go right through the dark matter distribution, which is much more extended and dispersed than the luminous baryonic component. The cluster dark matter mass acts as a "lens", which by deflecting the light inwards focuses it and magnifies it. This has two effects, one of which is to intensify the light (making it easier to detect distant faint objects thus lensed), and the other being the formation of distinct lensed images offset by an angle from the true direction towards the source. The phenomenon is complicated and depending on the impact parameter between the ray and the center of the DM distribution, as well as the shape of the DM distribution, can result in multiple images of the same object, obtained from light going through

Figure 4.3 Gravitational lensing of background sources by the foreground cluster Abell 2218, showing various arches of partial Einstein rings.
Source: W. Couch, HST-NASA.

different paths.[1] In the case when the source and the center of mass of the foreground cluster are perfectly aligned, the image has the shape of an "Einstein ring" of angular half-opening angle (in radians) $\theta_E = (2R_s D_{LS}/D_S D_L)^{1/2}$, where D_S and D_L are the distance from us to the source and us to the lens, and D_{LS} is the distance between the lens and the source. Such measurements allow one to map the distribution of, and the total mass of, the dark matter. An example of gravitational lensing by a foreground cluster of galaxies is shown in Fig. 4.3. Such observations provide one of the most important experimental measurements of the overabundance of DM relative to baryonic matter, confirming also the much smoother and extended distribution of the DM relative to the luminous mass.

The gravitational lensing measurements confirmed the original inference of the existence of DM in clusters (due to Zwicky), dating back to the 1930s. This was based on the measurement of the dispersion of the velocities of galaxies in galaxy clusters of radius R. From Kepler's law $V = (2GM/R)^{1/2}$, this gives the amount of gravitating mass M inside the radius R, which is about a factor 8–10 larger than that of the observed luminous (baryonic) mass. The evidence for dark matter in galaxy halos dates from the 1960s, based on the "rotation

---

[1] Since different paths are involved, the light arrives with different time delays, which become noticeable in the case of time-variable sources. The path lengths are proportional to the Universal distance scale $1/H_0$, so such measurements provide a method to determine Hubble's constant H_0.

curves" of galaxies measured by Vera Rubin and others. The observed rotation velocities of stars and gas are measured as a function of their distance from the center of rotation of the galaxy. This, from Kepler's law $V(r) = (2GM(r)/r)^{1/2}$, gives the amount of gravitating mass $M(r)$ at each radius r, which leads to the findings that (a) the DM exceeds considerably the amount of visible baryonic (stars and gas) mass, and (b) the invisible DM responsible for the rotation velocities is distributed in a much larger halo than the luminous baryonic matter.

In the LCDM scenario, star formation starts to occur soon after the halos have turned around and started to recollapse. This is thought to give rise to the presently observed galactic stellar halo component of "old" stars. The most naive expectation is that if the DM halo and the gas did not have much initial angular momentum and if the gas supply was exhausted through star formation during this initial infall, the resulting galaxy would be a gas-poor elliptical galaxy (see Fig. 4.4). On the other hand, if the gas was endowed with enough angular momentum, and if star formation did not proceed too fast, there would be a significant fraction of gas which forms a flat rotating disk, where star formation continues. Spiral arm patterns can emerge in such disks as the result of interactions or mergers with other galaxies or smaller satellites, leading to spirals similar to our own (Fig. 4.5). In all cases the rotating gas disk's outer radius is substantially smaller than the radius of the virialized parent dark matter halo. The real story is likely to be more complicated, and in particular mergers

Figure 4.4   The massive elliptical galaxy Messier 87 (M87), located in the nearby Virgo cluster of galaxies. A smaller spiral galaxy in the same cluster is also seen near edge-on.
Source: Canada–France–Hawaii Telescope & Coelum.

Figure 4.5 The spiral galaxy known as Messier 81, or M81, in a composite image from NASA's Spitzer and Hubble space telescopes and NASA's Galaxy Evolution Explorer. It shows elegant spiral arms which curl all the way down into its galactic nucleus. It is located about 12 million light years away in the Ursa Major constellation and is one of the brightest galaxies that can be seen from Earth through telescopes. Our own galaxy, the Milky Way, is also a spiral which would look similar if viewed from outside. The nucleus of M81, the Milky Way and other galaxies are known to harbor a massive black hole of millions of solar masses and sometimes larger.
Source: NASA.

are likely to play a significant role in stripping gas from merging halos and in changing the angular momentum. The above simple scenario at any rate provides a basic description of the situation, which can be tested through numerical simulations such as those shown in Fig. 4.2.

During the collapse of the dark matter and gas structures, the central density of the collapsing structure naturally increases. The orbits of the dark matter particles and any stars formed during the radial infall result in higher densities near the center of the proto-galaxy. Also, gravitational scattering of stars by each other as well as gravitational friction between stars leads to dissipation of orbital energy leading to a further mass concentration towards the center, which results in the formation of a *galactic nucleus* where the stellar density is much larger than in the rest of the galaxy. Stellar collisions lead to massive stars which quickly evolve into black holes (BH), as discussed below. Larger black holes are formed through the continued collisions and mergers of stars among themselves and with their black hole remnants, eventually resulting in the formation of a central massive black hole (MBH), which can grow to encompass millions to billions of solar masses. It is also possible that the very first stars to form were extremely massive, due to the low primordial abundance of cooling heavy elements, and these would have collapsed very early into black holes of mass in the range of hundreds or perhaps up to a thousand solar masses, giving a head start to the formation of massive black holes. It is not yet clear whether these MBHs formed early on and played a significant role in the subsequent development of the proto-galaxies, or whether they developed along the way, as the proto-galaxies developed, fed by the infall of gas, disrupted stars and whole stars which drifted in towards the center.

4.3   Stars: the Universe's worker bees

Much of what we know about the Universe and our local neighborhood is derived from observations of the luminous stellar components, and to a lesser degree of the gaseous component. However, it is important to realize that these are only the luminous *markers* of the distribution of the bulk of the dark matter mass, which dominates the overall gravitational field and the dynamics. Stars are also the progenitors of black holes, which are even more luminous markers, thanks to the radiation of the gas falling into them. Black holes are the most concentrated form of mass-energy in the Universe, acting as *focal points* where the most energetic and violent processes in the Universe occur. The stellar-sized black holes, which may be called "dead" stars or at any rate their remnants, are the end product of the evolution of the more massive "normal" stars, which despite this somewhat bourgeois appellation are in themselves also quite magnificent objects.

Stars are the most abundant form of mass-energy concentrations. Their typical internal mass densities $\rho_* \sim 1\,\mathrm{g\,cm^{-3}}$ are enormously larger than the averaged densities of their host galaxies, the latter being (including dark matter) equivalent to about ten atoms per cubic centimeter, $\rho_{gal} \sim 10^{-24}\,\mathrm{g\,cm^{-3}}$. The contrast is even larger between the density inside stars and the average densities of present clusters of galaxies, which is about $10^{-27}\,\mathrm{g\,cm^{-3}}$, and even more so when compared to the average mass density of the present Universe, $\rho_U \sim 10^{-29}\,\mathrm{g\,cm^{-3}}$. Stars are the foot-soldiers of the Universe. Although terribly commonplace in a universal sense, they are extremely impressive on a human or even planetary scale, with masses typically millions or tens of millions of times larger than the Earth mass, and with large luminosities generated by nuclear fusion reactions in their inner core, which is absent in planets. The typical main sequence stellar mass is in order of magnitude comparable to the mass of the Sun, $M_\odot = 2 \times 10^{33}\,\mathrm{g}$, the average main sequence stellar densities of order $1\,\mathrm{g\,cm^{-3}}$ being comparable to that of water or earthly solids. The stellar masses range approximately from 10^{-1} to $\lesssim 10^2 M_\odot$, and the stellar luminosities range over several orders of magnitude around that of the Sun, which is $L_\odot = 4 \times 10^{33}\,\mathrm{erg\,s^{-1}} = 4 \times 10^{26}\,\mathrm{W}$.

Stars, including our Sun, derive their luminosity from thermonuclear reactions in the stellar core. This involves mainly the fusion of protons into deuterium, and eventually into helium, etc. The luminosity of the Sun is equivalent to the explosion of 100 trillion (10^{11}) thermonuclear (fusion) bombs per second, rated at 1 megatons of TNT each. By comparison, the energy consumption rate of the entire human population in 2005 was $\sim$ 15 TW (terawatts) $= 1.5 \times 10^{13}$ W. This is comparable to the power delivered by one small fission bomb of 4 kilotons of TNT exploding every second, involving the fissioning of 0.3 kg of U-235 per second, or 10 000 tons of U-235 per year. Instead of using explosions, nuclear fission reactors are a safer way to currently provide 16% of this power usage to the world. To provide the full 15 TW of world power usage would require a sixfold increase in the number or capacity of nuclear reactors, requiring about 15 000 tons of U-235 per year, extracted from 3 million tons of naturally occurring uranium per year and enriched at 4.4%. The conservatively estimated assured world reserves of U-235 in the Earth's crust amount to about 3 million tons [24], sufficient at the present rate for hundreds to thousands of years, depending on the degree of recycling. Sea water is another natural source of uranium, whose reserve is estimated at 4.5 billion tons, or 1500 times the assured ground reserves. The other alternative is provided by controlled fusion reactions. Fusion reactors have been under investigation for decades, but a commercially viable version still remains to be achieved.

The Sun, of course, is a highly viable fusion reactor on its own. The problem, if we are thinking of ambitious industrial applications, is how to utilize it. If only one part in a trillion (10^{-12}) of the luminosity emitted by the Sun over 4π were captured and converted with 10% efficiency into electricity, one would in principle be able to supply 3000 times the power consumption of the Earth.[2]

---

[2] Of course, the problem is intercepting a sufficiently large fraction of the solar power emitted into 4π. The Earth as seen from the Sun subtends a center to edge half-angle of 8.7 seconds of arc, or a solid angle of 5.7×10^{-9} steradians, so the solar illumination at the top of our Earth's atmosphere over the equator at noon is 0.136 W cm^{-2}. Due to absorption in the atmosphere, clouds, non-equatorial location, etc., this illumination is reduced by about a third on average at ground level, being equivalent to a 100 W light bulb placed 5 inches above the ground. The whole illuminated hemisphere intercepts about 5×10^4 TW, so if ideally 1% of the surface of the Earth were covered with 10% conversion efficiency photo-cells, this could supply 50 TW, or three times the current Earth power consumption. Still, a number of practical factors have been ignored here, which would considerably reduce the above efficiency. For example, 1% of the Earth's

However, even aside from human hopes for future exploitation of the Sun, solar illumination of the Earth has had enormous and beneficial consequences. Life owes its origin to solar heat and irradiation, and the growth of higher species, agriculture and civilizations is directly attributable to it. More generally, nucleosynthesis inside stars (the fusion of lighter elements into heavier ones, starting from hydrogen) is responsible for the production of elements such as carbon, nitrogen and oxygen which make up most organic matter, as well as silicon and iron group elements which make up much of the Earth. The even heavier elements such as gold, titanium, uranium and others are in turn made in the supernova explosions which are the end point of the evolution of massive stars. Without stars the observable world would be quite different, and possibly even lifeless.

4.4   Stellar and galactic concentrates

Brilliant as they are, stars pale into insignificance compared to what they become after they can no longer support nuclear fusion. Indeed, the luminosities and mass densities of their successors, stellar remnants such as white dwarfs, neutron stars and stellar mass black holes, are many orders of magnitude larger than those of their former selves. These compact remnants are left over after the stars have died, so to speak, having exhausted their nuclear fuel supply. This leads to a collapse of their core, which is accompanied by the ejection of their outer envelopes in the spectacular display of a planetary nebula or a supernova remnant. Figure 4.6 shows a diffuse remnant called the Crab nebula, which is left over from the outer layers of a star which underwent a supernova explosion, leaving a compact neutron star remnant near the center (the Crab pulsar), discussed further in Chapter 6.

The most extreme mass-energy concentrations associated with the deaths of stars are the gamma-ray bursts. These objects are thought to be associated with the formation of compact remnants which are probably stellar mass black holes, as discussed in Chapter 7. GRBs are the largest and most intense explosions known in the Universe, shining mainly in gamma-rays, and delivering themselves of about a thousandth of a solar rest-mass worth of photon energy in a matter of seconds. In other words, in a few seconds a GRB pours out an amount of electromagnetic radiation energy comparable to the entire light output of the

total area is $\sim 2.5 \times 10^8$ acres $\sim 10^8$ hectares, a considerable loss of usable land area; then there is the cost of producing the huge area of photo-cells, which would require also a considerable initial use of power; the cost of installing power lines to conduct the electricity, and so on.

Figure 4.6   The Crab nebula supernova remnant, resulting from a star which self-destructed about 1000 years ago in 1054 AD. Near its center is a compact remnant, in this case a neutron star. Other supernovae resulting from more massive progenitor stars lead to black hole compact remnants.
Source: NASA, ESA.

Sun over its entire lifetime of $\sim 10^{10}$ years, i.e., about the age of the Universe since the Big Bang occurred. It is also comparable to the entire light output of our Milky Way galaxy (composed of about 10^{10} stars) over a year. And that is without taking into account an expected, even larger but so far unobserved energy output from GRB in gravitational waves, neutrinos and cosmic rays. This additional energy output is estimated as up to a thousand times larger than the electromagnetic output, or a fraction of a solar rest-mass in gravitational waves and thermal neutrinos, while the cosmic ray and non-thermal neutrino output may lie somewhere between the output in the form of electromagnetic and gravitational waves.

The heavy-weights among mass-energy concentrations are of course the massive black holes found at the center of galaxies, with masses ranging from millions to billions of solar masses. These massive black holes, as mentioned, grow to such impressively large sizes by disrupting and swallowing neighboring stars and gas, as well as through mergers with other black holes, aggregating into successively larger mass black holes. As more and more stars, gas and smaller black holes are sucked into the growing central black hole, the latter plays the macabre role of a growing cemetery cum compactor of dead stars. This cannibalism is atoned for by the splendid commemorative display of intense radiation issued by the MBH as it gorges itself (Chapter 5). If stars may be compared to worker bees, black holes, especially MBHs, may be compared to queen bees, just sitting there, receiving food, and glowing in splendor. The sequence of accretion and mergers is enabled by viscous friction, instabilities, tidal dissipation and gravitational wave emission, all of which sap the orbital angular momentum of the surrounding gas and stars, causing them to spiral down the thus greased gravitational funnel into the yawning abyss of the black hole (*facilis descensus Averni*, in the insightful words of Virgil and Dante). Despite being a highly schematized scenario, we can see how this can lead to such large MBHs as observed.

The mergers of galaxies lead not only to a more massive successor galaxy but also, as it appears in many cases, to the merger of their central black holes, which thus become even more massive. Initially the two MBHs would become bound in orbit around each other, and drift closer by emission of gravitational waves (see Chapter 8). The typical frequency of the gravitational waves in mergers of $10^6 M_\odot$ MBHs is of the order of one wave crest per hour or so, in the millihertz range, comparable to the orbiting frequency of the two black holes around each other as they merge. The range of sensitivity of the Laser Interferometer Space Antenna (LISA), a planned ESA/NASA gravitational wave detector in space (see Chapter 9) is $10^{-1}-10^{-4}$ Hz, which will be most sensitive to mergers of MBHs in the mass range $10^3-10^7 M_\odot$.

Depending on the mass of the MBH as well as the gas and stellar density in the immediate environment of the MBH, which gets larger in the case of galaxy mergers, the luminous energy output of the MBH ranges from rather modest values up to $L \sim 10^{48}$ erg s^{-1} $\sim 10^{15} L_\odot$ or more. These photon luminosities are fueled by the gravitational accretion of gas around the MBH, as well as accretion of stars in their neighborhood which are gravitationally captured and disrupted before being swallowed. The more modest MBHs (for instance the $M \sim 3 \times 10^6 M_\odot$ MBH at the center of our Milky Way galaxy) have luminosities not much above $L_{BH} \sim 10^{38}$ erg s^{-1} $\sim 10^5 L_\odot$, comparable to that of many stellar mass black hole or neutron star accreting binaries. In such galaxies, which are by far the most numerous, the central black holes are only very weakly radiative, due to a low accretion rate, and thus are hard to detect directly. They are more easily detected by indirect means, through their gravitational effects on the orbits of nearby stars, or the concentration of their diffuse light.

However, about 1% of galaxies harbor MBHs of mass $M_{BH} \sim 10^7-10^8 M_\odot$ in their nuclei, in which the activity is much more spectacular. These are called active galactic nuclei (AGNs), and include objects such as Seyfert galaxies and quasars. In these, the accreted gas drifting down towards the MBH horizon gets heated to millions of degrees or more, and radiates photons ranging from the ultraviolet to the X-ray range, in a more or less continuous although gently varying manner, with luminosities ranging from 10^{45} to $10^{47}-10^{48}$ erg s^{-1}, i.e., up to $L_{AGN} \sim 10^{15} L_{\odot}$. There are also even less common but more extreme AGNs, whose radiation is predominantly very non-thermal. This emission is generally associated with jets of low density gas arising from the inner parts (near the MBH horizon) of the accreting flow, which are ejected with relativistic velocities. These include the radio-loud quasars and radio galaxies, which are discussed further in Chapter 5.

4.5   Black hole characteristics

Violent events associated with black holes occur in the Universe on many different scales. The primary energy source in all of these events is gravity. The fact that the basic energy source is the same results, not surprisingly, in observational properties and in particular radiation properties which are also similar across the large range of mass scales. This similarity is somewhat reminiscent of the scaling behavior described by fractal theory, where the system properties are self-similar across a large range of scales. In the present case, the scale is set by the mass of the black holes involved. For a non-rotating black hole, or one which is not spinning too fast, the fundamental length scale is the size of the light horizon, called the Schwarzschild radius R_S. Its value is

$$R_S = 2GM/c^2 \simeq 3 \times 10^5 (M/M_{\odot}) \, \text{cm}, \tag{4.2}$$

which increases linearly with the mass.

The Schwarzschild radius is strictly derived from General Relativity, but some appreciation for its meaning can be gained from a quasi-heuristic Newtonian argument due to John Mitchell in 1783 and the Marquis de Laplace in 1796. Consider a mass M which lies inside a radius R. A test particle of small mass m located at a very large distance r will be attracted by M and will start to fall towards it at the free-fall velocity $v = (2GM/r)^{1/2}$. As m approaches M the free-fall velocity increases, and if M is sufficiently compact, the radius of approach r could in principle become small enough that the velocity of the test mass m could reach the speed of light c. This is so far purely Newtonian mechanics. However, if we dial fast forward to the 20th century, we know that no physical speed can exceed the speed of light. Thus the limiting radius at which the particle m would reach the speed of light is $r_{lim} = (2GM/c^2) = R_S$.[3] This is the previously mentioned

---

[3] During its infall, and still within a Newtonian approximation, at each radial distance r the test particle has a kinetic energy $(1/2)mv^2$ and a gravitational energy $-GMm/r$. If

Schwarzschild radius, inside which the mass M resides. Thus, naively (but relativistically correctly) one concludes that the mass m crosses the radius $r = R_S$ at the speed of light.

Furthermore, the test mass m could send out during its infall light signals which travel at the speed of light c, and these could reach a distant observer as long as the test mass was at distances $r \geq R_S$. However, once the mass crosses below $r = R_S$, no further signals can be received by a distant observer, since no signals can travel at a speed in excess of the speed of light. The Newtonian argument is here only formally indicative and cannot be entirely relied upon, but it conveys an intuitive picture of the situation.

Another way to look at this is to imagine an astronaut jumping upwards. On the Earth, he can jump maybe a meter, at best. From a planet with the same mass M as the Earth but much larger radius R (smaller $2GM/R$), he could jump much higher. It is for this reason that on the Moon astronauts can bound several meters high, although in that case the mass M of the Moon is smaller than the Earth's but its radius is such that $2GM/R$ is smaller than on the Earth. On the other hand, on a planet of the same mass but much smaller radius than the Earth (much larger $2GM/R$) the astronaut would barely be able to jump a few centimeters. And if $2GM/R$ approaches c^2 (that is, R approaches the Schwarzschild radius), the astronaut would not be able to jump even a micrometer. Similarly, a rocket being fired off from a radius r with the aim of escaping the clutches of the mass M (the black hole) can only escape if it is ejected from a radius $r > R_S$, where the gravitational escape velocity (the free-fall velocity with an inverted sign) does not formally exceed the speed of light, $(2GM/r)^{1/2} < c$. Formally, from radii smaller than this, the rocket would have to be given a velocity in excess of c to escape, which is impossible. Thus, the meaning of the gravitational horizon or the light horizon of a mass M is that, if the mass M is located inside its own Schwarzschild radius $R_S = 2GM/c^2$ (that is, we are dealing with a black hole of mass M), no information, particles or even light can escape from within its horizon.

In General Relativity, space-time is distorted by the presence of a large mass, and test masses move along the natural slope of space-time, on paths called

the infall started from an approximately infinite radius, initially both the infall velocity (hence the kinetic energy) and the gravitational energy are zero, as is the total initial energy, which is the sum of the two: $E = (1/2)mv^2 - (GMm)/r = 0$. In the absence of friction, the total energy is conserved, and this remains so as the mass m approaches M. As the test particle approaches M, its velocity v and its kinetic energy $(1/2)mv^2$ grow, but this is canceled out by the increase in the gravitational energy $-GMm/r$. The mass m falls in at a "free-fall" velocity, given from the conservation of energy as $v = (2GM/r)^{1/2}$. Requiring that $v \leq c$ leads to the limiting Schwarzschild radius $r \equiv R_s = 2GM/c^2$.

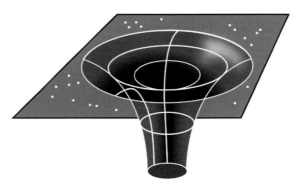

Figure 4.7   Schematic illustration of the distortion of space-time around a black hole, leading to the "Schwarzschild throat" at the bottom of the diagram. *Source*: NASA.

"geodesics". A partial grid of these is represented in Fig. 4.7. There are radial geodesics, which converge towards the black hole, and test masses which far away had no angular momentum fall radially along such radial geodesics, eventually crossing the Schwarzschild radius. Test masses which had some angular momentum settle into more oblique or transverse geodesics resembling the Newtonian circular or elliptical orbits. If the test masses in transverse orbits lose angular momentum, for instance due to friction, they move down into tighter orbits, until eventually they too can cross the Schwarzschild radius. Far away from the Schwarzschild radius, the radial and transverse geodesic paths are essentially coincident with the Newtonian orbits, but near the Schwarzschild radius there are substantial differences introduced by General Relativity. For instance, for radii r within a few Schwarzschild radii, the circumference ℓ of a circular orbit is no longer $\ell = 2\pi r$, having instead a value $\ell < 2\pi r$. The most striking difference, however, is the presence of a limiting radius, the Schwarzschild radius, where not only light but also massive test particles move at the speed of light [25].

It follows from the previous discussion that all observable phenomena associated with black holes are due to radiation emitted by matter which is at radii $R \geq R_S$, for instance light emitted by gas which is compressed and heated up as it falls towards the black hole. Similarly, any gravitational or any other type of radiation which may be detected must also arise from outside the Schwarzschild radius [26]. The accreted matter generally has some angular momentum, which leads the inflow to settle into a rotating accretion disk. Viscosity conveys the matter in the disk inwards down to a last stable orbit somewhat above the light horizon, from which it plunges into the BH in a free-fall time. For a Schwarzschild BH, or one which is not spinning too fast, the radius of the last stable orbit is at

$$r_l \equiv 3R_S = 6GM/c^2 = 0.9 \times 10^6 (M/M_\odot)\,\text{cm}. \tag{4.3}$$

This is the characteristic "observable" length scale associated with a black hole. For a solar mass black hole this is about 10 km, a distance which a human could walk in three hours at a good pace. The fractional gravitational binding energy at this orbit is 0.0572, which means that the maximum efficiency for conversion of gravitational energy into radiation of any kind is about 6%, for matter drifting inwards along the accretion disk around a Schwarzschild black hole.

Of course, when an initially non-rotating black hole accretes matter which is rotating, the black hole will acquire angular momentum and will start to spin up. Rotating black holes are called Kerr black holes, being represented by a different solution of the general relativistic equations. The last stable orbits in such Kerr black holes are either smaller or larger than the Schwarzschild last stable orbit for the same mass, depending on whether the disk or the test particle is orbiting in the same sense (prograde) or in the opposite sense (retrograde) as the black hole itself. It can be shown that there is a maximum possible rate of rotation for Kerr black holes, and for such maximally spinning black holes (also called extreme Kerr black holes), the last stable orbit in the equatorial plane is at $r_{l+} = GM/c^2 = 1.5 \times 10^5 (M/M_\odot)\,\text{cm}$ (prograde orbit), or at $r_{l-} = 9GM/c^2 = 1.35 \times 10^6 (M/M_\odot)\,\text{cm}$ (retrograde orbit). In the prograde case the maximum gravitational energy extraction efficiency is 42.26%, while in the retrograde case it is 3.77%.

The essential features of the accretion phenomena onto black holes are simplest to discuss in the non-spinning Schwarzschild case, which within factors of order unity gives a good estimate for most situations. The general relativistic value of the angular velocity of matter orbits accreting along a disk, including the last stable orbit, is essentially similar to the Newtonian or Keplerian value $\Omega = (GM/r^3)^{1/2}$. The corresponding orbital velocity is $v = (GM/r)^{1/2} = (c/\sqrt{2})(R_S/r)^{1/2}$, which is the dynamical velocity at that radius, and similar in magnitude to the free-fall velocity. The Keplerian orbital time around the last stable orbit r_l, which is also essentially the free-fall time from that radius, is

$$t_l \simeq 2\pi r_l/v_l = 4.6 \times 10^{-4}(M/M_\odot)\,\text{s}. \tag{4.4}$$

Thus, r_l and t_l are the (minimum) characteristic length and time scales associated with black holes of mass M. For a solar mass black hole, this is about half a millisecond, the time needed for light to go around a 10 km radius circle;

the orbital frequency is about 2 kHz, which is about half the frequency of the highest note on a piano, and roughly 10 times higher than the frequency of typical neuronal signals in our brains.

4.6   Black hole astrophysics

Based on the predominant type of radiation emitted, black holes can be classified broadly into two families: those where ultra-energetic photon and particle emission dominates, and those where gravitational effects and gravitational wave (GW) emission dominates. The photon (and particle) loud family BH sources have strong mass monopole or dipole moments, meaning isotropic or axially symmetric motions, but they lack a strong mass quadrupole, meaning motions in along one axis and out along a perpendicular axis. On the other hand, the gravity wave loud family sources have large mass quadrupoles, with either comparable or weaker mass monopole and dipole moments. Both families range from stellar to supergalactic scale, and black holes of different masses play a major role in both. Within these two families a further sub-classification can be based on the level of violence with which they emit these radiations. Aside from details, the overall behavior and the stages through which these violent events progress are remarkably similar on all scales, along both families.

Among the photon-loud sources, on stellar scales the most violent events are also the most intense explosions known in the Universe, the gamma-ray burst sources. As mentioned, these are related to the formation of a compact central remnant which is (or quickly becomes) a black hole of several solar masses, resulting in the ejection of a highly relativistic jet which emits very high energy radiation. These events also liberate a huge amount of energy in the form of prompt thermal neutrinos, of energy approximately tens of megaelectronvolts, amounting to a fraction of a solar rest mass (where $M_\odot c^2 \sim 2 \times 10^{54}(M/M_\odot)$ erg) over a time of about ten seconds. A relatively smaller, but still stupendous amount of energy of a few times 10^{51} erg, of the order of percents of a solar rest-mass, emerges over timescales of seconds to tens of seconds in the form of gamma-rays from a jet which broke free from the stellar debris. One sub-class of GRBs, the so-called short GRBs, are probably much more strongly GW loud than they are photon loud. On the other hand, the more photon-loud "long" GRBs are thought to be miserly GW emitters, compared to their own electromagnetic and particle energy output and also compared to short GRBs.

On galaxy scales, the most violent photon loud (and probably also particle loud) sources are those present at the center of a small percentage of galaxies, the active galactic nuclei (AGN), powered by central massive black holes of

$10^7-10^9 M_\odot$. These MBHs accrete the surrounding gas and stars, resulting in a relativistic jet of galactic scale which emits larger fluxes of ultra-high energy radiation. The intensity and duration of the emission episodes roughly scales with the BH mass, but it depends also on the accretion rate.

In both stellar and galactic photon loud sources, rotation and angular momentum are thought to play a prominent role in allowing a jet to be formed and to escape from the clutches of the gravity of the black hole. These sources produce comparatively little GW emission.

In the family of the predominantly GW loud sources, on stellar mass scales the pride of place belongs to the compact binary merger sources, where the binary system consists of two neutron stars (NS–NS), a neutron star and a stellar mass black hole (NS–BH), or perhaps two stellar mass black holes (BH–BH). The loss of orbital angular momentum due to the accumulating effects of gravitational wave emission leads eventually to the merger of the two compact objects, leading to a burst of GWs during the final plunge, leaving behind a more massive black hole. The energy in GWs thus emitted is of the order of a solar restmass emitted in milliseconds. The objects are tentatively identified also as the "short" class of GRBs, very luminous in gamma-rays and probably also in ultra-high energy particles, but still only at the level of parts in a thousand of their expected GW luminosity. On galaxy scales, all normal galaxies are thought to house at least a moderately massive black hole at their center. Our Milky Way, for instance, appears to have a $3 \times 10^6 M_\odot$ MBH at its center, which affects the orbits of the stars in its neighborhood. The relatively low level of electromagnetic (i.e., photon) nuclear activity is thought to be due to a much slower rate of mass accretion into such "normal" galaxies, or possibly also a slower spin rate of the black hole. Occasionally, however, a star will wander too close to the MBH even in such normal galaxies, and this should lead to a burst of gravitational waves, liberating a fraction of the rest-mass of the disrupted star over a timescale comparable to a near-grazing orbital time of $t_g \sim 300(M/10^7 M_\odot)$ s. This will be accompanied by an electromagnetic flare in the optical and X-ray from the heated gas of portions of the disrupted star as it is being swallowed.

Also in the GW loud family, between the galactic and the stellar mass scales there are probably many intermediate mass black holes residing in the globular stellar clusters, which populate the halos and the galactic bulges of galaxies. These are also likely to be too leisurely rotators to make jets, but as with their larger galactic center cousins, they are expected to be sporadic sources of gravitational waves as they accrete newer stars during their growth.

On the much larger scales of clusters of galaxies, occasionally two galaxies approach close enough to undergo a gravitational merger. As the two MBHs in their nuclei merge, they emit a powerful burst of gravitational waves, of the order of a fraction of the rest-mass of the two MBHs over a grazing orbital time. This enormous amount of energy will also be accompanied by an electromagnetic (optical and X-ray) flare as the resulting MBH captures and swallows the disturbed stars and gas in its neighborhood. These electromagnetic flares, as well as those from stellar captures by MBHs in single normal galaxies, represent only a very small fraction of the energy of the GW flare itself, but cheaper and more efficient photon detectors make them easier to detect than the more powerful gravitational waves.

5

Active galaxies

5.1   What makes a galaxy "active"?

Galaxies range in mass from $\lesssim 10^9 M_\odot$ to $\gtrsim 10^{12} M_\odot$, and are broadly classified into spiral, elliptical and irregular types. Irregulars appear only in the lower mass range, while galaxies with total masses above $\sim 10^{10} M_\odot$ are usually spirals or ellipticals. Examples of spirals are our own Milky Way galaxy, or the galaxy M81 shown in Fig. 4.5. Ellipticals exist over the whole mass range, but the most massive galaxies are typically ellipticals. Above a total mass of roughly $5 \times 10^{10} M_\odot$ galaxies show a more or less developed nucleus; that is, a concentration of stars, gas and dark matter which forms a noticeable bulge at the center, which is generally much brighter than the rest of the body of the galaxy.

One of the most notable discoveries of the last three decades is that those galaxies which have a substantial nucleus appear to have a massive black hole at its center, which plays a significant role in the evolution of the nuclear region. Thus, our Milky Way galaxy has a black hole of $M \sim 3 \times 10^6 M_\odot$. Some galaxies appear to have black holes several orders of magnitude larger than that. The black hole mass is usually inferred from the dynamical motions of the stars and gas in its neighborhood, where the mass of the black hole dominates the gravitational potential. In most galaxies, the central black hole does not lead to a significant increase in the luminosity of the galactic nuclear region. There are several possible reasons for this. One of these may be that the capture of gas and stars by the black hole is infrequent, as in the case of our galaxy, due to a density of stars in the nucleus which, while large compared to the solar neighborhood, is still too low for more than very infrequent captures. Another reason for an apparent nuclear passivity is that the central black hole can be so large that the tidal forces acting on the stars being swallowed are not able to disrupt them, swallowing them whole instead, without any large luminosity flare – no squash, no splash.

In a small fraction of galaxies, however, on the order of 1–2% of the total, the nuclear regions do show an unusually increased luminosity relative to the rest of the galaxy. In these, the nucleus is active: gas is being accreted in large quantities and/or stars are being captured, disrupted and swallowed in significant numbers. This leads to an extremely bright, and sometimes highly variable nucleus, whose luminosity can exceed by many orders of magnitude the luminosity output of the stabler stellar population in the body of the galaxy.

Such galactic nuclei are called AGNs (active galactic nuclei), and colloquially the host galaxies themselves which harbor such active nuclei are often referred to as AGNs, since compared to the nucleus the brightness of the rest of the galaxy pales by comparison, to the point that from far away often only the nucleus is detected. The typical *nuclear* luminosity of a moderately bright AGN is

$$L_{AGN} \simeq 10^{13}L_\odot \simeq 4 \times 10^{46} \, \text{erg s}^{-1} \sim (1M_\odot \, \text{yr}^{-1}) \times c^2, \tag{5.1}$$

while the luminosity of the rest of the host galaxy is typically $(10^{11}-10^{12})L_\odot$. The nuclear luminosity often varies by factors of two or more, $\Delta L \gtrsim L$ over timescales of $\Delta t \sim$ hours or less, which allows one to infer that the size of the nuclear emitting region cannot be larger than

$$r \lesssim c\Delta t \sim 10^{13} \, \text{cm}. \tag{5.2}$$

This is because such large changes of luminosity must involve the emission from the whole emitting region, and the difference in the arrival time to the observer of the increased light from the front and the back of the emitting region gives a lower limit for the observed variation timescale. This estimate assumes that the emitting gas in the vicinity of the central object is moving with at most semi-relativistic speeds. The inferred size must thus be on the order of or less than the Schwarzschild radius $R_S = 2GM/c^2$ of a black hole of mass $M \sim 3 \times 10^7 M_\odot$ [27].

The luminosity must be due to mass accretion of an amount of mass ΔM over a timescale Δt. For accretion on a black hole, this process can convert gravitational potential energy into radiation with a typical efficiency of $\eta_r = (L/c^2)(\Delta t/\Delta M) \lesssim 0.1$, as discussed in Chapter 4. The luminosities (5.1) imply that the corresponding accretion rates are of order $\sim 10 M_\odot \, \mathrm{yr}^{-1}$. The matter will be highly ionized by the intense radiation field, so the main effect of the radiation on the accreting matter will be due to scattering of photons by the free electrons (the scattering by the protons is negligible, being smaller by the ratio of the square of their masses $(m_e/m_p)^2 \sim 3 \times 10^{-6}$). Thus, the light pushes the electrons outwards, but the protons cannot be left behind, because this would cause enormous electrostatic forces between regions of unbalanced negative and positive electric charge, which would end up causing them to move together. While the force of gravity is trying to pull them inwards, at the free-fall velocity if unhindered by radiation, the radiation pressure is trying to push the matter outwards, and the higher the accretion rate the higher is the luminosity. The possibility exists that a very high accretion rate could lead to a luminosity which stops the accretion. This occurs when the two forces are equal, at a critical luminosity called the Eddington luminosity, whose value is[1] $L_{Ed} = 10^{38} (M_{BH}/M_\odot) \, \mathrm{erg \, s}^{-1} = 1.25 \times 10^{46} (M_{BH}/10^8 M_\odot) \, \mathrm{erg \, s}^{-1}$. Taking an average gravitational energy conversion efficiency $\eta_r \sim 0.1$ and requiring that the mass accretion rate $\Delta M / \Delta t$ should not produce a luminosity so high as to stop the accretion defines similarly a maximum Eddington mass accretion rate, $(\Delta M / \Delta t)_{Ed} = (L_{Ed}/\eta_r c^2) = 1.4 \times 10^{18} (M_{BH}/M_\odot) \, \mathrm{g \, s}^{-1} = 2.1 (M_{BH}/10^8 M_\odot) \, M_\odot \, \mathrm{yr}^{-1}$. The Eddington rate increases proportionally to the mass of the MBH. Thus, the AGNs with the highest observed luminosities are expected to have MBH masses of up to $M_{BH} \sim 10^9 - 10^{10} M_\odot$.

AGNs have been surveyed extensively at all wavelengths and out to some of the highest redshifts observed. The properties related to the black hole luminosity have been mapped in deep surveys at X-ray wavelengths [28], providing markers for the evolution of compact structures at redshifts up to the reionization epoch, where they are the brightest steady state photon sources.

5.2   MBH masses, masers and distances

All AGNs are thought to contain a central massive black hole, but evidence for the value of the mass is for the most part in the form of lower limits, based on the observed nuclear luminosity L, and applying the Eddington restriction. This is an approximate estimate at best.

---

[1] The electron scattering cross-section is $\sigma_T = 6.6 \times 10^{-25} \, \mathrm{cm}^2$. The radiation pressure $f_r = (L/4\pi r^2)(\sigma_T/m_p c)$ exerted on the plasma (including the mass of the protons) must not exceed the gravitational force $f_g = G M_{BH}/r^2$. Thus the luminosity should not exceed the Eddington luminosity $L \leq L_{Ed} = (4\pi G M_{BH} m_p c / \sigma_T) = 1.25 \times 10^{38} (M_{BH}/M_\odot) \, \mathrm{erg \, s}^{-1} = 1.25 \times 10^{46} M_8 \, \mathrm{erg \, s}^{-1}$. This is for spherical symmetry, whereas in reality accretion often occurs along a disk, but within numerical factors the results apply for disks as well.

However, in a limited number of AGNs and galaxies, there are more direct methods of probing the mass of the central massive black hole. For instance, in our own Milky Way galaxy, which does not harbor an AGN, there is evidence for a black hole of mass $M_{MW} = 3 \times 10^6 M_\odot$, which is derived from mapping over the years the motions of individual stars around the central nuclear source. Such observations [29, 30] are done in the infrared, in order to penetrate through the high dust column density along the galactic plane which obscures the optical view.

These have provided for a number of individual stars both transverse (through imaging) and longitudinal (through Doppler shifts) velocity components. Since the distance to the galactic center is known (8.1 kpc), the transverse angular separations translate into known transverse distances to the gravitating center (the MBH), which statistically must be comparable to the longitudinal distances, so that the radial distance of the stars to the gravitating center is known, and simple gravitational dynamics yields the black hole mass. For instance, in the case of simple circular orbits, the square of the Keplerian orbital velocity (or the freefall velocity) is

$$v^2 = GM_{BH}/r, \tag{5.3}$$

from which, knowing the radius and the velocity from observation, one deduces the mass. This result is easily generalized to elliptical or more general orbits. This method (see Fig. 5.1) has yielded among others the mass of our own Milky Way central black hole, $M_{MW} = 3 \times 10^6 M_\odot$ [29,30]. Of course, this method cannot be used on more distant galaxies, since the apparent angular sizes are much smaller, but for nearby galaxies it can be done in a statistical way. For example, one way is by measuring radial Doppler velocity dispersions as a function of angular distance from the maximum light concentration, and other related techniques have been applied as well. For more distant objects, as is the case for many AGNs, this gets difficult.

An especially important case occurs in some AGNs hosted in spiral galaxies which are seen not too far from edge on. Spiral galaxies, our own included, have dust and gas, including in the inner few parsecs rings of dense gas rich in molecules. These molecular regions along the inner disk contain, among other compounds, water molecules, H_2O. In some AGN-hosting spiral galaxies, including some so-called Seyfert 2 galaxies, the nucleus produces intense continuum radio emission, and this radio emission can excite the water molecules to radiate as a maser at very narrow, well-defined frequencies. Since the maser frequencies are so well defined, any changes of frequency due to mass motions are easy to detect via the Doppler shift of the frequencies, and this allows us to measure

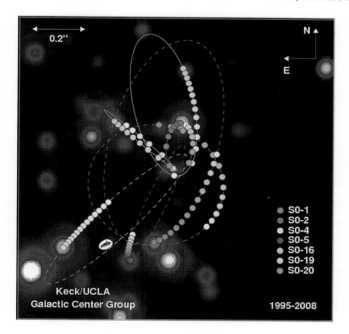

Figure 5.1   Orbits of stars around our Milky Way central black hole mapped in infrared light as a function of time [30].
Source: These images were created by Professor Andrea Ghez and her research team at UCLA and are from data sets obtained with the W. M. Keck Telescopes.

the gas motions near the nucleus. These motions are of course dominated by the central black hole, so the dynamics of the gas give a direct measure of the black hole mass. An example is the Seyfert galaxy NGC 4258 [31], see Fig. 5.2.

Radio measurements with the Very Long Baseline Array (VLBA) radio interferometer achieve milliarcsecond angular resolutions easily, and are able to map the water maser lines along the galactic disk, not only in this Seyfert galaxy but also in a number of other AGNs seen close to edge on. Because the maser emission arises from such a large amount of gas, these sources are often referred to as *Megamasers*. Since the galaxy is seen close to edge on, the core radio continuum source is observed directly shining through the disk, which contains the molecular ring. It also illuminates the rest of the ring, including the outermost approaching and receding edges. The line of sight velocities are given by $V_r = \sqrt{GM/R}\sin\theta$, where R is the radius of the ring and M is the mass of the MBH. These velocities can be plotted as a function of the angular distance in milliarcseconds (mas) from the center. A plot of the line of sight velocity of the maser lines as a function of distance from the nucleus is shown for another Megamaser galaxy, the Seyfert 2 UGC 3789, in Fig. 5.3. The two

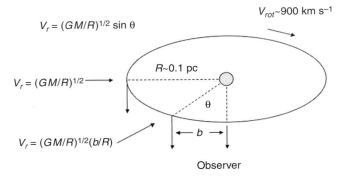

Figure 5.2   Schematic of the AGN NGC 4258 galaxy (after [31]). The central nucleus emits radio continuum which excites the maser emission of water molecules located in a ring seen at low inclination (the oval). The molecular ring of radius R is rotating with circular velocity $V \simeq (GM/R)^{1/2} \sim 900\ \mathrm{km\,s^{-1}}$. The line of sight component is multiplied by a $\sin\theta$ factor, which is unity (away or towards the observer) at the left or right extremes, and $\sin\theta \sim (b/R)$ for the gas moving almost transversely closest to the line of sight. Such measurements lead to a determination of the black hole mass (see text).

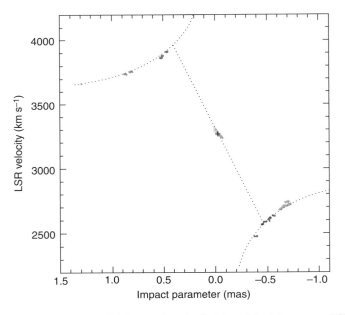

Figure 5.3   Line of sight rotational velocities of the Megamaser AGN UGC 3789 [32]. The abscissa is the transverse angular separation from the nucleus (proportional to the impact parameter b of Fig. 5.2).
Source: Reproduced by permission of the AAS.

curves on the left and right show the Keplerian behavior $V_r = \pm\sqrt{GM/r}$ of the receding and approaching portions of the ring. The central region shows the ring portion near the line of sight towards the core, where the velocity variation can be shown to depend linearly on the projected separation b. Measuring the radial and transverse velocities of individual gas blobs in the ring then gives the mass of the black hole.[2] Furthermore, since one measures the angular separation θ_R of the left and right inner edges of the Megamaser ring, one obtains the distance to the galaxy, $D = R/\theta_R$. For distant AGNs, this is in fact an exciting new way to measure their distances. This method is the basis for a key project to calibrate the Hubble constant H_0, the *Megamaser Cosmology Project* [32]. The Hubble constant relates the velocity of recession of the galaxies V_{gal} with their distance D. This Hubble relation, which at low redshifts is linear, $V_{gal} = H_0 D$, is a key quantity in cosmology. All our knowledge about the extent of the observable Universe, and how far different distant objects actually are, is based on a determination of H_0, which currently is known through various methods to $\pm 10\%$. The goal of the Megamaser Cosmology Project is to reduce this to $\pm 3\%$. An accurate value of H_0 would also constrain other important cosmological parameters, including the geometry of the Universe (whether it is open, closed or flat) and the fraction Ω_m of the critical density contributed by matter.

For distant AGNs, say with $z \gtrsim 2$, the above methods become difficult. By far the most widely used method for measuring MBH masses at high redshifts, extending now to $z \sim 6$, is the so-called *reverberation mapping* method [33]. This works for AGNs where broad line regions are well observable (most quasars and Seyferts, but unfortunately not blazars). The bulk of the luminosity from the quasar is continuum radiation from the nucleus, and this excites the atomic lines of the broad line region (BLR), which is located at some distance R_{BLR} from the central black hole. At this distance the motion of the line-producing gas is still dominated by the gravity of the central massive black hole, so the frequencies of the lines are broadened by the Doppler effect. The central continuum luminosity generally varies in time, and this induces a change in the line luminosity, which however is delayed by a time $\Delta t \simeq c/R_{BLR}$, which is measured, thus giving R_{BLR}. The measured line frequency dispersion due to the

---

[2] Close to the line of sight $\sin\theta = b/R$, where b is the impact parameter and $V_r = b\sqrt{GM/R}$, which gives a linear relation between the line of sight velocity and the impact parameter (central portion of Fig. 5.3). It is possible to follow individual gas blobs as they accelerate through this midpoint, which gives the centripetal acceleration $a \simeq V^2/R$ directly. Since V is measured at the two edges, this yields a *physical* value of the ring radius R directly, even without knowing the distance to the galaxy. Since $V(R)$ is measured at the left and right inner edges of the ring, the mass of the black hole is $M_{BH} = V^2 R/G$.

Doppler effect directly reflects the velocity dispersion of the line-emitting gas, $\Delta v/v \simeq \Delta V/c$, which being dominated by the gravity of the central black hole is $\Delta V \simeq (GM_{BH}/R_{BLR})^{1/2}$. Thus, we obtain the mass of the central black hole as $M_{BH} \simeq fR_{BLR}(\Delta V)^2/G$, where f is a numerical factor of order unity which depends on the geometry of the BLR region. Sampling a number of different atomic lines at different distances results in a fairly accurate estimate of the central MBH mass, to within factors of 3–5. This has been done with a large number of distant AGNs, and has been very helpful in surveying the MBH populations as a function of redshift. It has also shown that the maximum MBH masses are about $10^{10}M_\odot$, and that extremely massive MBHs already exist at redshifts of $z \sim 6$.

5.3   An AGN garden, classified

AGNs are broadly classified as radio quiet and radio loud, their relative numbers being approximately ten to one. The former have little radio emission, which is generally concentrated in the galaxy core. The latter have strong radio emission, some of it in the core but most prominently in extended lobes of diffuse radio emission at the end of two opposed jets whose dimension often exceeds that of the optical galaxy by an order of magnitude or more.

The radio quiets are mainly (but not exclusively) in spiral galaxies, and they do not have prominent radio lobes, although some of them may have smaller jets of sub-galactic dimensions which are occulted by gas or dust. Sub-types include "LINERS" or weak Seyfert 2 types; Seyfert 2 (Sy 2), with weak optical continuum and narrow emission lines of Doppler widths $\lesssim 500 \, \mathrm{km \, s^{-1}}$, arising in "narrow line regions" (NLR) [34]; Seyfert 1 (Sy 1) with strong continuum and broad emission lines of Doppler widths $\gtrsim 500 \, \mathrm{km \, s^{-1}}$ arising in "broad line regions" (BLR); and radio-quiet quasars, which are 1–2 orders of magnitude brighter than Seyfert 2s.

Radio-loud AGNs generally have prominent radio lobes, which in some cases resemble narrow jets while in others they are more diffuse, but in both cases rather extended structures. Some of the strong sources are hosted in large elliptical galaxies, while some are in spiral galaxies. Sub-types include radiogalaxies, some of which have narrow lines (NLRG) while others have broad lines (BLRG), and the brightest ones are radio-loud quasars. Besides a jet which emits non-thermal radiation over a broad range of frequencies, they also have conspicuous nuclear emission, which was initially detected optically but which is also prominent over a wide range of frequencies, and in fact very often produces more energy at X-rays or gamma-rays. The radio jet morphology serves to classify radio-loud AGNS into so-called Fanaroff–Riley (FR) classes. The FR 1 class have less bright, fuzzy radio lobes and the nuclear radio emission is stronger than that of the jets, which are usually wide and asymmetrical. The FR 2 class are overall brighter in radio, and have long and narrow jets which are much brighter in radio than the nucleus.

The morphology of radio-loud and radio-quiet AGNS can be understood in terms of a simplified paradigm, called the *unified AGN* model. This assumes that all AGNs consist of a more or less flattened stellar distribution, which in the case of spirals includes a dusty gas torus or disk, and a jet which may or may not be present, emerging more or less perpendicular to the torus plane. The classification of such a generic AGN as one or the other type of AGN mentioned previously is assumed to depend largely on the orientation of the torus and the jet (if present) relative to the observer, and on the strength of the accretion of the central MBH (Fig. 5.4). When the observer's line of sight is at large angles relative to the jet, and/or when the torus is seen more or less along the plane, atomic lines are observed from gas clouds in and above the torus which participate in the galactic rotation. They are sampled across a wide range of azimuthal separations $r_\perp$ from the rotation axis, most of them being at large $r_\perp$, so their radial velocity dispersion $\sqrt{2GM_{BH}/r_\perp}$ is relatively small.

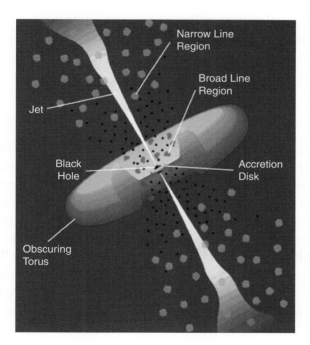

Figure 5.4 Schematic unified classification scheme of AGNs.
Source: NASA.

When the AGN is observed at intermediate angles, the line of sight samples more clouds which are further out and at closer distances $r_\perp$ from the axis, so their velocity dispersion $\sqrt{2GM_{BH}/r_\perp}$ is larger. This is a consistent, even if somewhat oversimplified interpretation of the narrow lines of radio-quiet Seyfert 2 and radio-loud NLRG, and of the broader lines of the radio-quiet Seyfert 1 and radio-loud BLRG.

The more extreme AGNs can also be interpreted broadly in this scheme as being those objects which are observed nearly perpendicular to the plane, or nearly along the jet axis. If the object is radio quiet (which generally goes with the absence of a jet), one is observing the nuclear region unhindered by any spiral arm gas or dust, and one observes the inner accretion flow onto the central massive black hole. For a typical inner accretion radius of $r \sim 3 \times 10^{13}$ cm and a luminosity $L \sim 10^{46}$ erg s^{-1}, one can estimate the typical photon temperature, since the disk will be dense enough for the radiation to be approximately thermal. Its spectrum at each radius resembles that of a black body, the luminosity from each radial ring being proportional to the fourth power of the disk temperature at that radius, $L(r) \propto r^2 T^4(r)$. The integrated disk luminosity is dominated by the inner part of the disk, since the disk temperature increases inwards faster than $r^{-1/2}$. Thus the peak photon energy from the integrated disk emission is $\hbar\omega \propto (L/r^2)^{1/4}$. Measuring the luminosity and estimating the inner radius of the disk (a small multiple of the Schwarzschild radius for a BH mass corresponding to a luminosity which is a fraction of the Eddington value), the characteristic peak photon energy of the AGN nucleus is

$$\hbar\omega \sim 3kT_{inner\ disk} \sim 100\,\mathrm{eV}\left(\frac{L}{10^{46}\,\mathrm{erg\,s^{-1}}}\right)^{1/4}. \tag{5.4}$$

Since the disk temperature decreases with radius as a power law, the added contributions of the black-body spectra at increasing radii end up resembling a non-thermal power law, which extends from the ultraviolet (UV) to the optical. The spectra of Seyfert galaxies are dominated in the UV/blue part of the spectrum by the disk. At cosmological distances, the redshift reduces this spectrum to a prominent "blue bump" in the optical spectrum, extending as a power law to longer wavelengths.

While the main mass of the accreting gas is in a dense flat accretion disk which produces the above UV/optical power law continuum, the disk is also expected to be surrounded by a much less dense outer layer, or corona, much hotter than the disk itself, which resembles the hot corona above the comparatively cooler solar atmosphere. The disk corona has temperatures which are in the $kT \sim 1$–30 keV range, and the hot coronal electrons scatter the cooler disk photons up to higher energies which reach up to the maximum energy of the electrons, $\hbar\omega \sim kT \sim 10-200$ keV. Quasars are indeed major extragalactic X-ray sources, and essentially all of the observed X-ray flux observed throughout the Universe can be resolved into the contributions from different quasars and AGNs distributed at various redshifts.

In the radio-loud AGNs the jets can reach distances of hundreds of kiloparsecs, tens of times larger than the size of the host galaxy. This is easier to determine for those AGNs which are observed at large angles to the rotation axis, since the large jets can be imaged in the radio with high angular resolution interferometric arrays, such as the VLA. The jets are typically "fed" at a variable rate, by the intermittent accretion rate onto their central MBH. Thus, the jets are prone to develop internal shocks within the jet, as gas parcels of different velocities hurtle outwards and run into each other. These are particularly prominent in the "inner" jet portions lying inside the galaxy, such as the famous luminous clumps seen in the inner jet of the nearby radio-galaxy M87 (see Fig. 1.2), which has been imaged in the optical, radio and X-rays. However, such internal shocks, as well as "re-collimation" shocks, are thought to be responsible for luminous structures seen in the outer jets extending tens of kiloparsecs outside the galaxy, as in the famous radio-quasar 3C 273. The jet is clearly seen in an X-ray image taken by the Chandra satellite, Fig. 5.5. In

Figure 5.5 Relativistic jet from the quasar 3C 273 in a Chandra X-ray image. The jet is seen also in radio, optical and gamma-rays.
Source: Chandra team, NASA.

addition to the internal shocks, termination shocks also occur at the outer end of the jets where they run into the intergalactic gas and are decelerated by its back-pressure.

In these various shocks, electrons (and protons) get accelerated by being repeatedly scattered back and forth across the shock interface, receiving a "kick" at each crossing which boosts their energy. This process is known as Fermi acceleration, and it results in the particles acquiring a power-law energy distribution, where the individual particle energies become extremely relativistic, with electrons reaching GeV and TeV energies. The magnetic fields in the shocked regions cause the accelerated electrons or positrons to radiate by the synchrotron process, resulting in a power-law photon spectrum extending from the radio to, in many cases, X-rays. These synchrotron photons as well as other ambient photons can also be scattered by relativistic electrons in the same or other parts of the jet, producing gamma-rays which in some cases reach GeV to TeV energies.

The bulk motion of the jet as a whole is in many cases itself significantly relativistic. This is quantified by the bulk Lorentz factor Γ defined (similarly to the individual particle Lorentz factor) as

$$\Gamma = \frac{1}{\sqrt{1 - (V_j/c)^2}}, \qquad (5.5)$$

where V_j is the jet bulk velocity and $\Gamma \geq 1$. Thus, as $V_j \to c$ the bulk Lorentz factor Γ gets much larger than unity. While in many older AGNs which have passed their peak activity period the outer jet velocities are sub-relativistic, $V_j \ll c$ and $\Gamma \simeq 1$, in some of the more extreme AGNs the jet velocity is very close to c and the bulk Lorentz factor has values of Γ up to 10 to 30.

In the case of highly relativistic jets which are pointing close to the line of sight to the observer, it is possible to observe what *appears* to be expansion velocities in excess of the speed of light. Individual emitting blobs in jets which are at modest angles to the lines of sight (10–15°) have been imaged in radio at different epochs, and their projected images on the sky appear to be moving apart at velocities which exceed the speed of light, $v_{app} > c$. Such apparently "superluminal" velocities can arise from jets or blobs which are moving at physical velocities $V \leq c$ but with high bulk Lorentz factors $\Gamma = [1 - (V/c)^2]^{-1/2} \gg 1$, see Fig. 5.6. This is purely a projection effect, and it does not violate the Special Relativity postulate stating that no physical object can travel faster than the speed of light.[3] The possibility of such apparent superluminal motions can be

---

[3] The light emitted from a blob at time t_1 at position 1 which is moving at a velocity v at an angle θ relative to the line of sight arrives at position 2 at time $t_2 = t_1 + \Delta t$ having

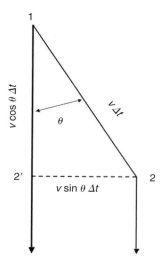

Figure 5.6   Generation of apparent superluminal velocity in an approaching object.

illustrated with an artificial example. Suppose that you shine a powerful laser beam onto a very distant screen, and describe an arc with your laser-wielding hand. If you swing your arm fast enough, and the screen is far enough away, the spot of light on the screen would appear at different positions at different times which, if it had involved a physical motion across the screen, would require speeds of transverse motion greater than c. However, no violation of relativity occurred: the laser light traveled at the speed of light, and your arm moved at much less than the speed of light.

The observation of such apparent superluminal velocities of blobs in radio AGNs seen at small angles has shown that in many cases the AGN jets have bulk Lorentz factors in the range of 10–30 or slightly more. The observations, however, are difficult for very small angles, being limited by the instrumental angular resolution. When the AGNs are viewed almost pole on, along lines

traveled a distance $v\Delta t$, where it is imaged again. The time interval for an observer between reception of the light ray from position 1 and that from position 2 is $(\Delta t)_{app} = \Delta t[1 - (v/c)\cos\theta]$ (Fig. 5.6). This is because from 1 the light signal traveled at speed c to a position 2' (referred to 2 by the dashed line perpendicular to the line of sight) while the blob traveled at speed $v \leq c$ to position 2. After that, from 2 and 2' the two light rays travel at the same speed. The apparent time delay is therefore $(\Delta t)_{app} < \Delta t$. The apparent transverse velocity in the sky is $v_{app} = [v\Delta t \sin\theta/(\Delta t)_{app}] = [v\sin\theta/1 - (v/c)\sin\theta]$. For some values of v and θ this can exceed c [35]. Differentiating v_{app} with respect to θ and equating to zero yields a critical angle $\sin\theta_c = [1 - (v/c)^2]^{1/2} = 1/\Gamma$, or $\theta_c \sim 1/\Gamma$ when $\Gamma \gg 1$, at which the maximum apparent superluminal velocity is obtained, $v_{app,max} = \Gamma v$.

of sight very close to the rotation axis, the radio-quiet QSOs (which have no
or weak jets) are seen mainly in the light of their accretion disk nuclear non-
thermal luminosity, while the radio-loud QSOs, which do have a jet, have this
jet pointing almost head on at us. For those AGNs where the jets are relativistic,
the Doppler effect boosts the intensity of the non-thermal radiation produced
by the jet, overwhelming the emission from the rest of the accretion disk and
the galaxy. These objects are called *blazars*, discussed in the next section and
also in Chapter 8.

5.4   Extreme AGNs

An extreme type of radio-loud AGN occurs when the jets are seen almost
head on, these being the above-mentioned *blazars*. Two variants of blazars are
known as BL Lac objects and flat spectrum radio quasars, or *FSRQ*, the latter
name indicating that the radio spectrum is a power law with a shallow slope.
Both of these sources are extremely bright, and generally highly variable non-
thermal gamma-ray sources. In fact, most of their observed luminosity is in very
high energy (VHE) gamma-rays, extending to the GeV–TeV range, as discussed
in greater detail in Chapter 8. These gamma-rays must emerge from beyond a
"gamma-ray photosphere" whose size is about $r \sim 0.1\,\mathrm{pc} \sim 10^{17}$ cm, otherwise
they would have been absorbed by interactions with other photons leading to
pair production, $\gamma\gamma \rightarrow e^+e^-$, and only photons softer than 0.5 MeV would have
been observed. The emission is highly variable, with luminosity changes of
order $\Delta L_\gamma \sim L_\gamma$ on timescales $\Delta t_\gamma \lesssim 1$ hr. Unlike the previous variability–size
relationship (5.2) for a non-relativistic source, the relationship for a source (in
this case the emitting jet) of dimension r which is expanding relativistically
with a bulk Lorentz factor Γ is[4]

$$r \lesssim c\Delta t \Gamma^2. \tag{5.6}$$

Thus, for observed inner jet dimensions of ~ 0.1 pc and variability times under
an hour, it is inferred that the jet bulk Lorentz factors must be of order
$\Gamma \sim 10$–30. These Lorentz factors lead to a strong Doppler boost of the jet
non-thermal continuum emission, which almost completely overwhelms any
line and thermal continuum from the host galaxy. Besides the non-thermal component associated with the nucleus or its jet, both these sources and quasars in
general also have a less luminous continuum radiation (UV or X-ray) associated

---

[4] Loosely, this can be understood by considering that the radiation comes from regions
separated by a distance r/Γ which was emitted over a time $\Delta t\Gamma$.

with an accretion disk, as well as emission line components of two types, associated with clouds further out. In the extreme blazars, however, the non-thermal jet emission beamed almost directly at the observer is so strongly boosted that its glare essentially "blinds" us and we are generally unable to detect the weaker disk and line emission. The two types of emission line regions are the so-called broad line regions (BLR), producing "permitted" atomic lines with $\Delta v/v \sim 0.05-0.1$ and Doppler widths 10^3-10^4 km s^{-1}, at radii $r \lesssim 10^{17}$ cm from the BH; and the narrow line regions (NLR), producing narrow permitted and forbidden lines of $\Delta v/v \sim 0.002-0.1$ and Doppler widths $\sim 50-500$ km s^{-1}, at $r \gtrsim$ pc (see Fig. 5.4). The line widths are naturally related to the Keplerian or dynamical (escape) times at their respective radii, being due to rotation, infall and/or radiation pressure driving. Both the disk continuum and the BLR are of interest as seed photon sources for Inverse Compton (IC) or photo-meson processes.

In blazars the fact that the jets are pointing close to the line of sight to the observer leads to the non-thermal jet radiation being both highly boosted and intermittent. The details of the gamma-ray phenomenology are determined both by the value of the jet Lorentz factor and by the angle at which the jet lies to the line of sight. Blazars typically have bulk Lorentz factors Γ in the range of 5–20, and most show a distinctive two-humped photon spectrum, the lower frequency hump being at optical to X-ray energies and the higher frequency one at GeV–TeV (10^{12} eV) energies. Both the optical/X-ray and the GeV/TeV emission in such blazars is episodic, occurring in flares which can last days, during which spikes of emission lasting minutes are observed. These flares alternate with longer periods of low-state "quiescent" emission, extending to lower photon energies in the hard X-ray to soft gamma-ray (MeV) range. The GeV–TeV emission of blazars and quasars is a very active subject, being pursued experimentally in the GeV range from space satellites such as Fermi and AGILE, and at TeV energies with various air Cherenkov telescopes. AGNs emitting in this very interesting energy range are discussed further in Section 8.3.

There are interesting parallels between the emission of GRBs (see Chapter 7) and blazars. The GRBs are powered by much smaller, stellar mass black holes, but the observed gamma-ray radiation is modeled in a manner similar to that in AGNs. What is striking is that despite their much smaller mass and geometrical size, the instantaneous luminosity of the average GRB in its prompt (outburst) phase is up to 10^5 times brighter than that of the brightest AGNs, albeit only for a very short time. The reason for this is that GRBs are a one-time catastrophe: their black holes form at the time of the event, and they convert a significant fraction of their total available rest-mass energy into a brief flash of high energy radiation. AGNs, on the other hand, while much more massive, accrete matter which is a smaller fraction of their own mass over longer times, and radiate away part of this energy at a comparatively less hurried pace. They do this, however, over a much longer period of time, so their time-integrated energy output is much larger than that of a GRB.

6

Stellar cataclysms

6.1 Stellar high energy sources

"Stars that go bang in the night" could be the title of a pervasive pop-song, and in fact the event is astonishingly frequent: it happens several times per second somewhere in the Universe, if you know where to look. The most common "bangs" are supernova explosions, which happen roughly every 100 years in an average galaxy. There are about 10^{10} galaxies within a Hubble light horizon, so the rate within our observable Universe is about 10^8 per year, or 3 per second. Supernovae and their remnants are, over a limited period of time, stellar high energy sources. In astrophysics we call high energy sources those objects which either emit non-thermal photon or particle radiation, or else whose emission spectra are thermal or quasi-black-body and emit mainly X-rays or gamma-rays. Typically these objects involve relativistic particle velocities. We have encountered galactic-size high energy sources in the previous chapter, and here we will discuss some of the scaled-down, stellar-sized counterparts of those high energy sources. Both the stellar and the galactic-scale sources are further sub-classified, not too accurately, as of high energy (HE) when the photon emission is up to the keV–MeV range, very high energy (VHE) when the emission extends to the $\sim$ GeV– TeV range, or ultra-high energy (UHE) when it extends much above the TeV range. For most sources it is generally the electromagnetic radiation which is observed, but the VHE–UHE qualifiers are also used for neutrino and cosmic-ray sources.

The high energy sources of stellar origin are of two types, compact and diffuse. The compact ones are generally stellar objects which are the rem-nants of the core regions of the original progenitor star. These remnant cores, which were dense to begin with, undergo further compression and cooling after nuclear reactions stop, which is accompanied by the rest of the stellar envelope being blown away. The high density and relatively low internal temperature of the resulting compact remnants leads to their pressure being provided mainly by the degeneracy pressure of the fermions in the remnant.[1] This is the case for white dwarfs and neutron stars, and black holes as well, although in the latter objects the pressure is inside the Schwarzschild radius and thus does not play a role as far as the outer manifestations of the remnant. These compact objects are the endpoints of the stellar evolution of stars with masses $M_* \gtrsim 3M_\odot$ [36]. The white dwarfs are soft X-ray emitters for a short time after they form, while white dwarfs which are in binary systems do so for longer times, which qualifies them as high energy sources. The neutron stars and black holes, on the other hand, while generally also seen in X-rays, often qualify as VHE or UHE sources, detectable in gamma-rays which can reach the GeV–TeV range. The supernova explosions which lead to the formation of neutron stars and black holes[2] also result in the ejection of the stellar envelope, which leads to a diffuse supernova remnant, or SNR. These diffuse remnants are themselves also high energy, VHE or UHE sources, since they accelerate particles to relativistic energies which emit non-thermal radiation.

6.2   White dwarfs and thermonuclear supernovae

6.2.1   White dwarf formation

The least degenerate of the high energy stellar sources, which have achieved so to speak only the first degree of degeneracy, are the white dwarfs. These compact stars arise from progenitor stars whose initial mass is in the approximate range of $\gtrsim 3-8M_\odot$, which have undergone nuclear fusion at their center for sufficiently long to have burned most of their initial hydrogen into heavier nuclei up to oxygen. In the process, the star passes from its main sequence stage to become a bloated giant star, with an extended and expanding outer envelope and an increasingly dense core made up of helium, carbon and oxygen. The core contraction proceeds until the electrons in it become

---

[1] The degeneracy pressure results from the combined effects of Heisenberg's uncertainty principle and Pauli's exclusion principle for fermions, which does not allow two fermions to approach closer than a minimal distance depending on the density.

[2] Supernovae associated with black holes are a phenomenon related to the gamma-ray burst sources discussed below.

degenerate, while the expanding envelope is blown off by the radiation from the core. Further contraction of the core is prevented by the electron degeneracy pressure, and in the absence of further compressional heating no further nuclear reactions are possible. The dense remnant core at this point has a mass of $M_{wd} \sim 0.5\text{--}1.4 M_\odot$, a radius of $R_{wd} \sim 10^9$ cm, a central temperature $T_c \sim 10^8\text{--}10^9$ K, and a surface temperature $T_s \sim 10^6\text{--}10^7$ K. It emits a nearly black-body spectrum, most photons having a mean energy $\hbar\omega_s \sim 2.8kT_s$ in the far UV to soft X-rays. These hard photons ionize the expanding envelope, which appears in photographs as a dramatic ring called a planetary nebula around the central remnant core. The Rayleigh–Jeans low frequency part of the latter's spectrum extends as $I_\nu \propto \nu^2$ down to optical wavelengths, where it appears to the naked eye as largely of white color, hence the name of White Dwarf. These are now essentially dead, or rather superannuated stars, which then just simply cool off by radiating away the thermal content that they previously accumulated, a process lasting in excess of the present age of the Universe, $\sim 10^{10}$ years.

6.2.2   White dwarf high energy sources

Some white dwarfs, however, come out of retirement and start on a second, much more flamboyant career, as high energy sources. This can happen especially to those white dwarfs which were originally in a binary system with another companion star, or those which at some point acquired a binary companion. This eventually leads to a transfer of mass from the companion, and accretion of gas onto the white dwarf (WD), which leads to two different and interesting outcomes. One of these is the phenomenon known as a *cataclysmic variable*. As the accreted gas falls onto the WD it forms an accretion disk, whose luminosity peaks in the far ultraviolet spectral range. The orbital rotation around the main sequence normal companion can lead, depending on the line of sight and the inclination of the orbit, to occultations of the WD by the companion, and variations in the mass accretion rate can lead to variability of the observed radiation. In some cases the WD has a magnetic field of $B \sim 10^6\text{--}10^7$ G, strong enough to channel the accreting matter onto the polar caps of the WD, where it gets shock heated to the virial temperature corresponding to the potential well $kT \sim (GM_{wd}m_p/2R_{wd}) \sim$ keV. As a result of the spinning of the WD around an axis other than the orbital axis, this leads to pulsing X-ray emission, which can be modulated by the accretion rate or undergo occultation by the disk and/or the companion. This is a class of widely studied X-ray binary sources, which are detected by space satellites such as Chandra, XMM, Suzaku, and others.

Some white dwarfs which are in a binary system go the extra mile to deserve the name of cataclysmic variables. They slowly capture and accrete matter from their companion, which leads to a low level of persistent UV/optical luminosity of the white dwarf. Eventually, enough accreted hydrogen and helium-rich material piles up on the surface of the WD so that its pressure becomes sufficient to ignite fusion. This "thermonuclear incident" leads, for a few weeks to a month, to a major flaring up of the white dwarf's luminosity, until the hydrogen is consumed. It does not, however, result in the disruption nor the collapse of the white dwarf, which then resumes accretion again until the next flare-up. This phenomenon is called a *nova*, not to be confused with the supernovae discussed below.

An even more violent fate is in store for some of the aging white dwarfs with obliging younger companions. This is caused by the increase in the total mass of the white dwarf due to matter accreted from the less evolved companion or, more rarely, due to merging with it, which leads to a significant increase of the white dwarf central temperature. As the mass approaches the so-called Chandrasekhar limit $M_{ch} \sim 1.38 M_\odot$, the core temperature becomes high enough to start carbon fusion, which ignites in a runaway fashion [36]. This is a thermonuclear explosion, called a *Type Ia supernova* (SN Ia), which releases an energy of $1-2 \times 10^{51}$ erg, sufficient to gravitationally unbind the white dwarf and completely disrupt it. The luminosity brightens suddenly to enormous values, rivaling the luminosity of the host galaxy itself (Fig. 6.1), making it highly

Figure 6.1   The Type Ia supernova SN 1994d, in the lower left corner of the galaxy NGC 4526, is almost as bright as the galactic nucleus, whose luminosity is $\sim 10^9 L_\odot$.
Source: NASA HST.

obvious to the naked eye if it occurs in our own galaxy, and easily detectable even if it occurs in other galaxies at cosmological distances. Because the core mass is always close to the Chandrasekhar value when they explode, the optical peak luminosity of this type of supernova is fairly constant, making them useful as "standard candles" for measuring cosmological distances, just as we can judge the distance at which we see a torch light of known intensity by the degree to which it appears fainter. This is an invaluable tool for tracing the dynamics of the expansion of the Universe. Its use has served to show that the expansion of the Universe appears to be accelerating, because the more distant supernovae appear fainter than they would be if the Universe had been expanding at the expected rate for the amount of known dark and visible matter in it. This has provided substantial evidence for the existence of dark energy as a major constituent of the Universe.

6.3   Core collapse supernovae

Core collapse supernovae are distinct from the thermonuclear supernovae discussed above, although in their outward manifestations they share similarities. They leave behind a different, more compact and degenerate type of stellar remnant, either a neutron star or a black hole. Core collapse supernovae (ccSNe) arise from progenitor stars with initial masses $M_* \gtrsim 8M_\odot$, which in the course of their evolution can burn carbon in their core, as well as the subsequent heavier elements they produce, including silicon, leading to the iron group elements (Fe, Co, Ni). In doing so, the outer envelope has expanded to about $10^3 R_\odot \sim 7 \times 10^{13}$ cm while the core has become very dense and compact, with a radius of $r_c \sim 10^9$ cm, roughly the size of a white dwarf. The iron group nuclei are the tightest bound of all, heavier nuclei being less bound, so that further nuclear fusion could only be exothermic; that is, they cannot proceed unless an external source provides energy. Lacking that, nuclear reactions naturally cease, and the core, deprived of its heat input, starts to cool, which in turn leads to loss of pressure support, resulting in the collapse of the core upon itself.

6.3.1   Core collapse and neutron stars

In these stars more massive than $M \gtrsim 8M_\odot$ the collapse cannot be stopped by electron degeneracy pressure, because the core has become more massive than the Chandrasekhar limit, and it is also hotter than in the less massive stars, so the electrons do not get degenerate until it is too late to stop the collapse. With the core thus falling upon its sword, the rest of the overlying star follows suit, and increasingly more distant layers start to fall in as well. As the core collapses, it starts to heat up as the density increases, which leads to two effects. One is that the photons become energetic enough to start dissociating the heavy nuclei synthesized in the previous evolution, undoing the previous work and leading back to a proton and neutron composition. The other is that as the density approaches values comparable to nuclear density, electrons start being absorbed by the protons, a process called inverse beta-decay,

$$e^- + p^+ \rightarrow n + \nu_e, \qquad (6.1)$$

which leads to "neutronization" of the core: most of the (by now hyperdense) core becomes made up of neutrons, with only a fraction (10% or so) of protons left over. The core thus becomes a proto-neutron star. The neutrons, at these very high densities, are degenerate, and the degeneracy pressure of the neutrons, if the core mass is less than $\sim 3-4 M_\odot$, is able to support the core against gravity.

This "stiffening" of the core puts a halt to the collapse of the overlying layers which smash onto it and get heated up to temperatures of $T_c \sim 10^{11}$ K. This leads to copious neutrino and antineutrino emission, which is initially trapped inside a "neutrinosphere" of about 30–50 km radius, in thermal equilibrium with the radiation and the matter. The increased pressure provided by neutrinos diffusing out of the neutrinosphere contributes to the reversal of the infall and the outward acceleration of the layers above the core, leading to a shock wave which propagates outwards through the star. Long before the shock wave reaches the surface, the thermal neutrinos escape, being the first harbingers of the supernova explosion. The total gravitational binding energy liberated by the collapse of the core to a neutron star is the difference between the gravitational energy at the beginning and at the end of the collapse,

$$\Delta E_B \simeq \frac{3}{5} \frac{GM_{NS}^2}{R_{NS}} - \frac{3}{5} \frac{GM_{NS}^2}{R_{core}} \simeq 3 \times 10^{53} \text{ erg} \simeq 2 \times 10^{59} \text{ MeV}, \qquad (6.2)$$

where the radius of the initial core $R_{core} \sim 10^3-10^4$ km is much larger than that of the final neutron star $R_{NS} \sim 10$ km, and $M_{NS} \sim 1.4 M_\odot$. A significant fraction of this liberated energy emerges as thermal neutrinos whose peak energies are in the 10–30 MeV range. Thus, the total output is on the order of $N \sim 10^{58}$ neutrinos, over a timescale of order ten seconds. Such a neutrino output was indeed observed from the supernova SN 1987a, which occurred in the Large Magellanic Cloud, at a distance of 51 kpc.

The visually most obvious leftover from the supernova explosion is the outer envelope of the star, which is ejected with a kinetic energy of about a percent of the liberated binding energy, $E_{kin} \simeq 10^{-2} E_B \sim$ few $\times 10^{51}$ erg. This energy is imparted by a shock wave, whose passage accelerates and heats the stellar envelope, which is ejected at speeds reaching $\sim 10^9$ cm s^{-1} or 10^4 km s^{-1}. An exciting observation in a recent supernova, SN 2008d, was what appears to have been the break-out of the shock wave, which was detected in X-rays as well as in ultraviolet light by the Swift satellite [37].

The ejected envelope becomes increasingly bright optically, until after a few days the photons finally become able to escape freely from it (the envelope becomes "optically thin"). This results in a first pulse of electromagnetic radiation, typically amounting to $E_{EM} \simeq 10^{-4}E_B \sim$ few $\times 10^{49}$ erg, which emerges a few days after the explosion. The electromagnetic radiation reaches a peak luminosity a week or so after, which is of the order of 10^{10} times the solar luminosity for a week or two. Since the number of stars of luminosity comparable to the Sun in a galaxy like ours is perhaps $\sim 10^{10}-10^{11}$, it is clear that for a few weeks the supernova competes with the whole galaxy in terms of optical luminosity. This stupendous optical display, visible to the naked eye when it occurs within our own galaxy, is a core collapse supernova. An example is given in Fig. 6.2, showing an image of the sky before (left) and after (right) a core collapse supernova.

6.3.2   Core collapse and black holes

A different stellar remnant ensues from the core collapse of even more massive progenitor stars, those with initial masses larger than $\sim 28-30 M_{\odot}$. In these, the mass of the collapsing core exceeds the theoretical stability limit

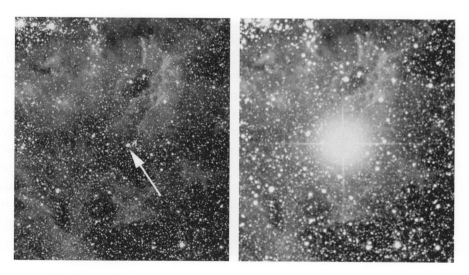

Figure 6.2   Two images of the same region of the sky before (left) and after (right) the explosion of the supernova SN 1987a.
Source: Anglo-Australian Telescope.

$M_{sl} \sim 3-4M_\odot$ for neutron stars, so the collapse cannot be halted and the core collapses through its gravitational horizon. This leads to a stellar mass black hole, the most compact and degenerate of the stellar mass remnants [38]. The collapse of the core may initially go through a temporary rotationally stabilized neutron star, whose fast rotation could lead to magnetic fields of order $B \sim 10^{14}-10^{15}$ G (magnetars), or perhaps "strange" stars in which strange quarks play a role. However, it is generally accepted that these intermediate stages can only temporarily halt the ultimate collapse to a black hole.

While the black hole outcome of the core collapse is widely accepted as a consequence of well-understood physics, the fate of the stellar envelope is less certain. A diffuse supernova remnant has been observed in a few well-documented cases associated with observed long gamma-ray bursts, which in turn are probably associated with stellar mass black holes (see Chapter 7). Gamma-ray bursts are observed in extragalactic objects only, due to their rarity, and the distance makes even a supernova explosion appear dim, which is one reason why only a few have been detected so far. The theoretical understanding of supernovae associated with black holes is on a poorer footing than that of the more frequently observed supernovae associated with neutron stars. While the numerical calculations of even the neutron star supernovae have failed, so far, to convincingly reproduce the ejection of the envelope through shock passage, at least the theoretical picture can rely on having an observationally well-documented central neutron star or pulsar, with relatively well-understood nuclear physics properties. The latter can provide degeneracy pressure support, as well as an intense if temporary source of neutrinos, both of which, qualitatively at least, appear able to turn around the collapse and power an outward accelerating shock wave. In the case of the core becoming a black hole, however, the degenerate pressure of the central object disappears and only the accretion disk is left to provide pressure and neutrinos. The need to take into account both general relativity and neutrino physics leads to a much more complex numerical problem, where in addition rotation effects in three dimensions play a large role. Thus, while efforts to model the situation numerically continue to make steady but slow progress, one is left for now with the observational evidence of the occasional diffuse supernova remnant, and the pious hope that the broad physical understanding of the physical principles behind such explosions is sufficient to describe the phenomenon at an approximate level of correctness.

6.3.3   *Diffuse supernova remnants*

Diffuse supernova remnants have been studied observationally for a long time, mainly in our own galaxy, where they occur at the rate of a few per century and where the relative proximity has made it possible to image them at different frequencies of the electromagnetic spectrum. This has allowed us

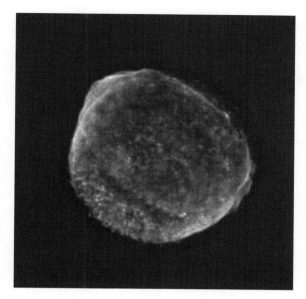

Figure 6.3 Supernova remnant SN 1006c, imaged by the Chandra satellite at two different X-ray energies.
Source: NASA Chandra team.

to follow the expansion of the envelope, showing that after reaching a luminosity peak a week or so after the explosion, the optical emission remains a brilliant, slowly fading display for thousands of years (the most famous example being the relatively young Crab nebula remnant shown in optical light in Fig. 4.6. Another example is the older supernova remnant SN 1006c, whose X-ray image is shown in Fig. 6.3. The radio, X-ray and gamma-ray emissions are other important components of the supernova remnant (SNR) emission. In particular, the interpretation of the TeV gamma-ray emission, detected in dozens of SNR with ground-based imaging air Cherenkov telescopes, is the subject of controversy. This has to do with SNRs being the most likely sources of cosmic rays at energies $\lesssim 10^{15}$ eV, whose diffuse flux is well observed, but which so far have not been directly associated with SNRs (see Chapter 10). One of the smoking gun proofs that SNRs are indeed cosmic-ray acceleration sites would be that these relativistic protons are expected to collide with thermal protons in the SNR leading to pions, whose π^0 component decays into gamma-rays in the TeV range. As discussed in Chapter 8, a controversy arises because an equally plausible mechanism for producing such gamma-rays is inverse Compton scattering of ambient microwave background and infrared photons by the relativistic electrons, which are responsible for the observed non-thermal X-ray emission of the SNR through synchrotron radiation.

6.4   Neutron stars and pulsars

Neutron stars are the other product of core collapse supernovae in stars of mass $M_* \lesssim 28 M_\odot$. After its initial neutrino burst and its decoupling from the expanding stellar envelope, the neutron star slowly cools over millions of years, and if it is not part of a binary system, it is detected sometimes in X-rays as an isolated, cooling NS, somewhere near the middle of a diffuse supernova remnant, provided the latter has not dissipated yet (which they do after $\sim 10^4 - 10^5$ yr).

More frequently, single neutron stars are detected through their regularly pulsating light curves at radio frequencies. These are the famous radio *pulsars*, first discovered in 1967 by Hewish and Bell. Neutron stars are initially endowed with strong magnetic fields, which in observed pulsars are inferred to be in the range of $B \sim 10^{11} - 10^{13}$ G. Such a magnetic field strength can in principle be simply understood in terms of conservation of the magnetic flux. The stellar material is almost completely ionized, and as such it is an excellent conductor, the magnetic field lines being "frozen" into the gas. As the gas gets compressed during the collapse, it drags the field lines with itself. Because of this flux freezing, and from the conservation of magnetic flux, the original stellar field $B_* \sim 1 - 10$ G is boosted during the collapse by a factor

$$B_{NS} \sim (R_*/R_{NS})^2 B_* \sim 10^{11} - 10^{13} \text{ G}. \tag{6.3}$$

There are, however, other mechanisms which could result in the even higher magnetic fields inferred in magnetars, discussed in Section 6.7.

The radio emission of pulsars is caused by electrons and positrons accelerated to extremely high Lorentz factors along the magnetic field lines, which beyond a few stellar radii are dipolar in shape (Fig. 6.4). The radio radiation is coherent, the electrons moving typically in bunches whose dimension is smaller than the wavelength of the emission they produce, and as a result the radio radiation is extremely intense. For pulsars whose magnetic axis is inclined relative to the rotation axis, a pulsating radio light curve is detected as the radiation beamed along the magnetic axis regularly sweeps the line of sight of the observer. Several thousand such radio pulsars are detected in our galaxy. The ultimate source of energy of a pulsar is the spin of the neutron star, since the accelerating electric potentials are proportional to powers of the spin rate and the magnetic field strength. As the spin rate or the magnetic field decreases, the pulsar emission weakens. Pulsars do show a slowing of the spin rate, which in combination with the period provides an estimate both of the age of the pulsar and of the magnetic field strength.

A few of the brightest pulsars detected, such as the pulsar in the Crab nebula, the Vela pulsar, etc., were formed recently enough (less than $10^3 - 10^4$ years ago)

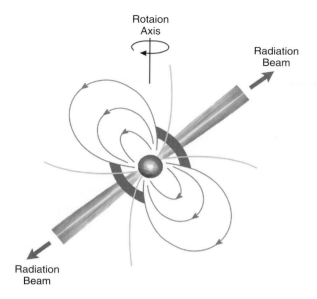

Rotaion
Axis

Radiation
Beam

Radiation
Beam

Figure 6.4   Sketch of a pulsar showing the dipole magnetic field and its orientation relative to the rotation axis and a pencil beam radiation directed towards the observer.
Source: NRAO.

that they can be observed to pulse also at optical, X-ray and even at gamma-ray wavelengths. The higher frequency pulsar radiation is thought to be due to inverse Compton scattering, whereby softer radio photons receive a boost in energy of order γ^2 as they collide with electrons of Lorentz factor γ. Pulsating gamma-ray radiation has been detected in the last year by the *Fermi* satellite at GeV energies from more than a dozen young pulsars (see Chapter 8). The energy source for this radiation is the rotation of the neutron star, which can be verified from the fact that the luminosity is observed to slowly decay as the spin rate decreases, i.e., as the rotation period lengthens. For this reason, these objects, which typically are not in binary systems, are called rotation-powered pulsars.

Some pulsars however are in binary systems with another pulsar, with a black hole, or with a white dwarf, but since they accrete either no or only negligible amounts of matter from these degenerate companions, their radiation still derives from their rotational energy. These binary pulsar systems are very interesting, because the regular variations of the pulse period due to the Doppler shift caused by the orbital motion allows extremely precise measurements of the orbital parameters and their changes. This has allowed, for example, determinations of the rate of change of the orbital separation caused by energy loss from the system due to gravitational wave emission, which gave the first experimental evidence for the existence of gravitational waves, as well as a confirmation of the general relativistic energy loss expression [39].

6.5   Accreting X-ray binaries

A substantial number of neutron stars which are not rotation-powered pulsars are found to be in binary systems with an ordinary star. Such binaries arise either because a supernova occurred in one member of a binary which was not unbound by the explosion, or because of capture into a binary system after the explosion. In these systems accretion from the companion onto the neutron star leads to intense X-ray emission, as the accreted gas gets heated to X-ray temperatures $kT \sim 10$ keV as it is decelerated and comes into approximate thermal equilibrium above the neutron star surface. Binaries where a compact object (either a neutron star or a black hole) accretes from a "normal" stellar companion are generally called X-ray binaries, and if the companion is a massive giant star the system is called a high mass X-ray binary (HMXB), or if it is a low mass companion it is called a low mass X-ray binary (LMXB).

The accretion flow, due to the orbital motion, forms an accretion disk around the neutron star, which seeps inwards due to viscosity, down to a radius which depends on the strength of the neutron star magnetic field. The initially strong magnetic field of neutron stars slowly dissipates with age, mainly due to conduction. In young enough systems, a magnetic field of order $B \sim 10^{11}-10^{12}$ G or stronger is able to stop the accretion disk at an *Alfvén* radius $r_A > R_{NS}$ where the magnetic stresses roughly equal the ram pressure of the accretion disk material,

$$\frac{B^2(r_A)}{8\pi} \simeq \frac{1}{2}\rho(r_A)v^2(r_A), \tag{6.4}$$

where $B(r_A) \simeq B_{NS}(R_{NS}/r_A)^3$ is the dipole field strength at r_A, and $\rho(r_A)$ is the accretion disk mass density at r_A (dependent on the accretion rate and the disk viscosity) while $v(r_A)$ is the disk Keplerian velocity $v^2 = GM_{NS}/r_A$ at r_A (Fig. 6.5).

In younger neutron stars which are in a binary system from which they accrete, the magnetic field is large enough that the Alfvén radius is significantly larger than the stellar surface. Upon reaching this radius the accreted matter cannot fall further inwards along the equatorial plane, but the matter can fall onto the star by following the field lines. Since the dipole magnetic field lines satisfy the relation $\sin^2 \theta/r = $ constant, it arrives at the neutron star surface inside a polar cap of radius $r_p = R_{NS} \sin \theta_p \sim R_{NS} (R_{NS}/r_A)^{1/2} < R_{NS}$. These are

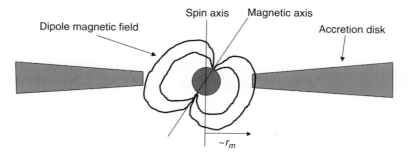

Figure 6.5 Sketch of an accreting X-ray pulsar showing the disk interface with the Alfvén surface at $r_m \equiv r_A$ which channels gas to fall onto the polar caps. *Source*: D. Chakrabarty, MIT.

"hot spots" on the rotating stellar surface, which again lead to a pulsating X-ray light curve. Since the energy source derives from the gravitational accretion of matter, these objects are called accretion-powered pulsars, or accreting X-ray pulsars (XRPs). Many dozens have been observed.

In older neutron stars, the magnetic field has decayed sufficiently so that the Alfvén radius is comparable to or less than the radius of the last stable orbit in the accretion disk, or the radius of the neutron star. In this case the accreted matter falls onto the equatorial regions of the neutron star where the disk grazes it, and the heated region spreads widely over the surface of the NS, leading to X-ray emission whose periodic variation is suppressed. These are essentially unpulsed accretion-powered neutron stars. In a few of these low magnetic field accreting neutron stars, the accumulated hydrogen-rich material can reach, after long enough accumulation, pressures sufficient to initiate thermonuclear fusion, giving rise to a very strong X-ray burst lasting for hours to days, with a distinctive light curve. These objects are known as X-ray bursters, being the neutron star analogs of the white dwarf novae.

6.6   Millisecond pulsars

While the large majority of the pulsating neutron stars detected either in radio or in X-rays appear to have fields in the range of 10^{11}–10^{13} G, there is a small minority of detected pulsars which have either much smaller or much larger fields.

The class of small field objects are almost always detected through radio observations, and their periods are generally very short, from milliseconds to tens of milliseconds, and for this reason they are known as millisecond pulsars. Their spin-down rates indicate values of the surface field of $B \sim 10^8$–10^9 G.

They are thought to be objects which initially had a higher field, which were or became part of a binary system. This can occur more easily for neutron stars inside dense groups of stars called globular clusters, where in fact most millisecond pulsars are found. The high density of stars can lead to occasional capture of a neutron star in a binary orbit around a normal star. Accretion of gas from the companion onto the neutron star could then have led both to a spinning up (thanks to acquiring high angular momentum matter from their companion) and at the same time to a suppression of the neutron star surface magnetic field. The latter can arise through a thermoelectric effect due to the fact that the outer surface is heated up by the infalling matter relative to the cooler inner core. This induces a current whose effect is to create a counterfield which diminishes the original field. Eventually the companion disappears, either through having lost so much mass as to destabilize itself, through becoming a supernova or through encounters with other stars disrupting the binary. The now widowed pulsar becomes then again a rotation-powered radio pulsar, which emits radiation at the expense of its rotational energy.

6.7   Magnetars

The class of high field objects is generally found through X-ray observations or through gamma-ray observations. Their very high fields are thought to arise during the initial core collapse, which results in very high temperatures leading to violent convective motions, which resemble the behavior of hot bubbles in boiling water. The material is of course highly ionized and the motion of electric charges leads to electric currents which provide a dynamo mechanism to generate magnetic fields, while the violent motion of the bubbles serves to stretch out and thus increase the strength of the seed magnetic field lines. If the rotation period of the proto-neutron star core is short, tens of milliseconds or less, the resulting field values can be very high, in the range of $B \sim 10^{14}-10^{16}$ G, and such objects are generically called magnetars.[3]

Anomalous X-ray pulsars.   One group of objects identified as magnetars are called anomalous X-ray pulsars, which has resulted in their acronym of AXP, where the very high field value is inferred (as most field values are) from their periods and their spin-down rate. Typically they are found to have periods in the range of 1–10 seconds, and fields in the range of $10^{13.5}-10^{14.5}$ G. The long period

---

[3] By contrast, the Sun's average surface magnetic field is 1 gauss, and the Earth's average surface magnetic field is 0.5 gauss. In terms of the more common industrial tesla unit, 1 gauss = 10^4 tesla. The large superconducting magnets which bend the particle beam around the LHC have fields of 8.33 tesla $\simeq 8.3 \times 10^4$ gauss.

can be ascribed to the fact that neutron stars born with tens of millisecond periods and such high fields would spin down extremely fast, but the spin-down rate decreases as the period lengthens, so that statistically one is much more likely to detect them as long period objects, since they evolve very quickly from their short period stage and spend most of their lives as long period objects. One of the interesting aspects of AXPs is that while in some of them accretion is occurring, there is evidence that a significant or even dominant fraction of their emission may derive from the dissipation of their magnetic energy. Based on the abundance statistics of normal and high field pulsars and the inferred lifetimes in each stage, it is estimated that up to 10% of all pulsars could have been born as magnetars.

Soft gamma repeaters. Another, even more exotic group of objects which are also identified as magnetars are the "soft gamma repeaters", or SGRs. These objects were initially thought to be full-fledged members of the gamma-ray burst family discussed in Chapter 7. However, they turned out to be a different type of animal, although there are similarities in their radiation which can make them look like short gamma-ray bursts. As the name indicates, these sources emit bursts of (soft) gamma-rays repeatedly, although irregularly, up to several times per year in some objects, and every few years in others (whereas true GRBs are considered to be one-time events). SGRs are now known to be *magnetars*, ultra-high magnetic field neutron stars whose fields are in the range of $B \sim 10^{13} - 10^{15}$ gauss whereas, as discussed in the previous chapter, the much more common pulsars have magnetic fields of $B \sim 10^{12}$ gauss.

It is thought that following the initial core collapse and rapid convective dynamo field build-up, after the debris has cleared somewhat, the extremely high field leads to rapid energy loss through emission of magnetic dipole radiation, which spins the magnetar down on a timescale much shorter than the time needed for the field to dissipate, so that (as also in the case of AXPs) most magnetars detected have long periods of the order of seconds. But unlike AXPs, SGRs flare up more abruptly and with most of their energy in soft gamma-rays, and their emission is much more fitfull. Some SGRs, as they continue to spin down, have been observed to undergo repeated but irregular episodes of burst-like gamma-ray emission. This is thought to be caused by a build-up of magnetic stresses in the neutron star crust, followed by a sudden realignment of the magnetic fields, perhaps due to cracking of the crust due to changing centrifugal stresses resulting from the spin-down. This causes a shaking of the foot of the magnetic field lines, causing a transverse perturbation of the field (called Alfvén waves) to propagate along the field lines, leading to acceleration of electrons and electron–positron pair formation, which results in a flare of gamma-rays and X-rays.

The flares sometimes reach very high intensities. The first example was the 1979 March 5 event in an object called SGR 0626-66 (the number indicates the galactic coordinates, that is, the location in a galactic map). It was eventually determined to be located in the Large Magellanic Cloud (known as the LMC), at about 50 kpc distance, and the gamma-ray light curve had an initial spike lasting a few seconds whose energy was $\sim 10^{44}$ erg. This is an extremely large energy for a neutron star, to be liberated in a short time, as can be appreciated from the fact that during those few seconds its luminosity exceeded that of the Sun by a factor of about 10^{10}. The clue that it was a neutron star was that this spike was followed by a much lower intensity tail of emission lasting well over 100 s, which showed very regular pulsations with an 8-s period, a tell-tale sign of a rotating neutron star.

Other SGRs were found, a total of eight as of 2009, located at "galactic" distances (i.e., either in our galaxy along the galactic plane or in very nearby galaxies such as the LMC). The most intense SGR by far was the 2004 December 27 giant flare of SGR 1806-20, whose initial spike had a luminosity of $\sim 4 \times 10^{46}$ erg and a pulsating tail of $\sim 10^{44}$ erg. Such luminosities, although extremely large compared to that of more steady galactic high energy sources (binary X-ray source luminosities do not exceed $10^{38}-10^{39}$ erg s^{-1}), do not allow SGRs to be detected much farther than the local group of galaxies, or perhaps the nearby Virgo cluster of galaxies (20 Mpc). Intriguingly, the initial $t \lesssim 2$ s spike of gamma-rays is harder than the subsequent weak long tail, and the spectrum of this initial spike is very similar to that of the so-called short GRB, which we discuss in the next chapter.

6.8   Stellar black holes

Most stellar mass black holes arise via one of two main channels. The first of these is the core collapse of stars so massive that the core mass exceeds the maximum mass which can be stabilized by neutron degeneracy pressure, so the collapse proceeds to a black hole final stage. The second channel is via matter accreting onto a neutron star from a binary companion star, until it exceeds the maximum mass limit for a neutron star and it collapses into a black hole. There are other possible channels, but these two probably account for the bulk of the stellar black holes in galaxies.

In the neutron star accretion scenario the ensuing black hole finds itself in orbit around a normal companion star, while in the core collapse scenario in a fraction of cases the collapsing star had another normal star companion. The anisotropy of the collapse and the envelope mass loss can lead to an unbinding of the binary leading to a single BH, but in a fraction of cases the binary appears to remain bound and the BH will eventually undergo further mass accretion. The mass of BHs in accreting binary systems is found to range from $\sim 3-4M_\odot$ (the maximum theoretical mass for stable neutron stars) up to roughly $10-15M_\odot$, the upper limit depending on the mass of the companion and on how much mass is lost from the system during the collapse process. The accretion then results in these BH binary systems being high energy radiation sources.

Most BH binary sources are detected as more or less steady X-ray sources. Depending on the orbital parameters of the binary, the X-ray emission may be modulated at the orbital period. This can occur regularly, for instance when occultation of the BH behind the normal stellar companion occurs, or when the accretion rate is modulated due to instabilities or due to orbital eccentricity where the mass accretion rate is modulated due to the regularly varying distance between the BH and its companion. In some cases the accretion ceases almost entirely except when the BH and the star are near the apogee of the orbit, that is their closest approach point.

Some accreting BHs sources, such as the source Cyg X-1, which has a blue super-giant stellar companion, vary erratically but are always detectable and do not undergo pronounced orbitally induced X-ray modulations. They do however show repetitive semi-regular variations of their X-ray luminosity, called quasi-periodic oscillations (QPO). These are probably related to variations of the minimum distance at which the accreting plasma approaches the black hole before it plunges in.

In addition, the X-ray emission of some BH sources such as Cyg X-1 alternates irregularly between two different modes, one where the X-ray luminosity is higher and the X-ray spectrum is softer (i.e., the maximum energy of the X-ray photons is lower, even if the total energy emitted is higher), and the other where the X-ray luminosity is lower but the spectrum is harder (i.e., the energy of the photons is higher). The hard spectrum is thought to be due to a low density hot corona above a thinner inner accretion disk, while the soft spectrum may arise when the inner disk puffs up and the corona cools off or is expelled.

Several BH binaries, besides their humdrum existence as steady or semi-steady X-ray sources, also occasionally undergo flaring episodes where they appear as microquasars (see Section 6.9). That is, they undergo flare-like episodes of increased accretion which result in the formation of a jet, which carries mass and energy at velocities which approach semi-relativistic values. An example is the X-ray source Cyg X-3, and also Cyg X-1, which for a long time were known just as semi-regular X-ray sources. The jets may be present at all times, but they become more obviously detectable only at times of increased accretion.

6.9   Micro-quasars: neutron stars or black holes?

In binaries where a compact object accretes from a "normal" stellar companion, when the compact object is a neutron star and accretion continues for a long enough time, the neutron star eventually must collapse to a black hole. Since the radius of a NS is only a few times larger than that of a BH of the same mass, the gravitational binding energy before and after differs similarly by factors of order unity, and the binary need generally not be disrupted. It is also possible that other black hole binaries are formed by capture of single black holes from tidal encounters with a normal star, in an environment dense enough for such encounters to occur.

These BH binaries will accrete, or continue to accrete from their companion, in what as before is called either a massive (HMXB) or a low mass (LMXB) X-ray binary. If the progenitor neutron star was not magnetized sufficiently to lead to detectable X-ray pulsations, the X-ray emission from accretion onto the black hole is not too different in its general properties from what it was when accretion occurred on the neutron star, since most of the radiation is generally due to the disk. One striking difference, though, is that low magnetic field neutron star binaries occasionally erupt in a thermonuclear X-ray burst from ignition of accumulated hydrogen-rich matter on its surface. Black holes, of course, have no such solid surface for matter to accumulate on, it just falls in once it reaches the last stable orbit in the disk. Hence X-ray bursts do not occur in BH-containing X-ray binaries. Neither do they show X-ray pulsations – the black hole "has no hair" (i.e., it does not have its own magnetic field attached to a firm rotating surface to channel accreting matter onto it, for the simple reason that nothing can stick out or exude from a light-horizon, other than its gravitational effects).

Besides their spectrum, another property that low magnetization NS and BH binaries appear to have in common is that of showing QPOs in their X-ray light curves. That is, the intensity shows increases at intervals of time which are almost but not quite repetitive – the intervals are fuzzy, sometimes shorter and sometimes longer, but roughly predictable. (By contrast, in either rotation-powered or accreting-magnetized neutron stars, the periodicity of the intensity increases is highly predictable, appearing as regular pulses which serve as very accurate clocks.) In low magnetization neutron stars and in (unmagnetized) black hole binaries, on the other hand, the quasi-periodicity may be due to effects associated with the rotation period or other periodicities of the accretion disk. For instance, either the inner edge of the disk changes, or hot-spots form in the disk, whose position is variable and only predictable on average. These are only two possibilities, there being a wider variety of models which attempt to explain this complex phenomenon. In these models, as in the observations, the general properties of the radiation are largely independent of the nature of the central object, depending mostly on its mass.

The most puzzling property found in micro-quasars and in a subset of X-ray binaries is that the accretion is variable, leading to alternating periods of the previously mentioned high luminosity/soft X-ray spectrum and low luminosity/hard X-ray spectrum. In addition to this, and more irregularly, the luminosity of micro-quasars occasionally flares up violently, which is connected with the appearance of a semi-relativistic jet emitting non-thermal radio as well as X-rays and gamma-ray radiation, the latter reaching into the GeV–TeV range. In these accreting binaries the light curves and Doppler measurements of the orbital velocities provide constraints on the orbital parameters, which in a few cases are compatible with a neutron star, but most often the mass is indicative of a black hole. It is for the latter reason that these sources are referred to as micro-quasars, resembling on a stellar scale the extragalactic massive black hole-powered jet sources producing non-thermal radiation. The non-thermal radio radiation is generally associated with the jets, and has provided, as for many radio galaxies and AGNs, evidence for apparent superluminal motions indicative of relativistic

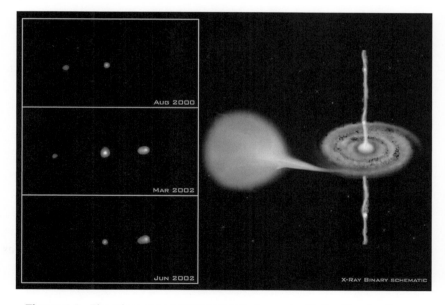

Figure 6.6   The micro-quasar XTE J1550-564 observed with the Chandra X-ray observatory showing (left) relativistically expanding jet components and (right) an artist's schematic conception of the object.
Source: NASA.

velocities in the jet. Relativistic motions have also been detected by means of X-ray observations. An example is the microquasar XTE J1550-564, where the jet motion was detected at various epochs with NASA's Chandra orbiting X-ray observatory (see Fig. 6.6). However, while in AGNs the bulk Lorentz factors are of order $\Gamma \sim 5-30$, in micro-quasars they are typically more modest values of $\Gamma \sim 1.05-1.1$, implying physical velocities of $v/c \simeq 1-1/2\Gamma^2 \simeq 0.4-0.7$. A number of micro-quasars have also been observed (see Chapter 8) at TeV gamma-ray energies by ground-based air Cherenkov telescopes such as HESS, MAGIC, CANGAROO, and VERITAS.

7

Gamma-ray bursts

7.1 What are gamma-ray bursts?

Gamma-ray bursts are sudden, intense flashes of gamma-rays, detected mainly in the MeV gamma-ray band. When they occur, for a few seconds they completely overwhelm every other gamma-ray source in the sky, including the Sun.

GRBs were first discovered in 1967 by the Vela military satellites, although a public announcement was only made in 1973. These spacecraft carried omnidirectional gamma-ray detectors, and were flown by the US Department of Defense to monitor for nuclear explosions which might violate the Nuclear Test Ban Treaty. When these mysterious gamma-ray flashes were first detected, it was determined that they did not come from the Earth's direction, and the first, quickly abandoned suspicion was that they might be the product of an advanced extraterrestrial civilization. However, it was soon realized that this was a new and extremely puzzling cosmic phenomenon. For the next 25 years, only these brief gamma-ray flashes were observed, which vanished quickly and left no traces, or so it seemed. Gamma-rays are notoriously hard to focus, so no sharp gamma-ray "images" exist to this day: the angular "error circle" within which the gamma-ray detectors can localize them is at best several degrees, which contains thousands of possible culprits. This mysterious phenomenon led to huge interest and to numerous conferences and publications, as well as to a proliferation of theories. In one famous review article at the 1975 Texas Symposium on Relativistic Astrophysics, no fewer than 100 different possible theoretical models of GRBs were listed, most of which could not be ruled out by the observations then available.

Until 1997 GRBs remained largely undetected at any wavelengths other than gamma-rays, aside from a few detections by the Japanese/US Ginga satellite in hard X-rays, with similarly rough spatial localizations. Gamma-ray observations however led to a first major breakthrough after the launch of the CGRO in 1991. This large NASA satellite had four different gamma-ray instruments, extending from sub-MeV to GeV energies. Especially with its omni-directional BATSE (Burst and Transient Spectroscopy Experiment), it provided such a large number of gamma-ray-based rough positions that it became obvious that GRBs were very isotropically distributed, and hence they must be either very close or else very far, at cosmological distances.

A dramatic improvement in understanding of these sources occurred in 1997 after the Italian–Dutch satellite Beppo-SAX succeeded in discovering longer duration X-ray afterglows of GRBs. These X-ray observations yielded several arc-minute accuracy positions after roughly one hour, refined to roughly one arc-minute several hours later. Taking into account the delays in processing and transmission to Earth, after about typically 8 hours these allowed ground-based telescopes to start searching for longer duration optical and radio counterparts. Such longer duration X-ray, optical and radio afterglows of GRBs were expected, having been predicted from theory before the observations were made. With the optical afterglow observations it became possible to identify the host galaxies, and thus measure their distances, which indeed turned out to be cosmological. After the demise of Beppo-SAX this task was continued by the HETE-2 satellite (2001).

More profound insights into the central engine and afterglow mechanisms were gained with the launch of the Swift satellite (2004), which threw light on a number of new hitherto unknown or only guessed properties of the prompt and afterglow multi-wavelength radiation, and made possible a rapid increase in the number of bursts with ground-based follow-up and distance determinations (Fig. 7.1). This was made possible thanks to two new capabilities of this satellite. The first of these was the greater sensitivity of Swift's burst alert detector (energy range 20–150 keV), while the second was its ability to rapidly slew to point at the burst its high angular resolution X-ray and UV/optical detectors. It is capable of doing this re-pointing ("slewing") in less than 100 seconds after the BAT triggers, yielding prompt and detailed multi-wavelength early afterglow spectra and light curves [40].

7.2   Phenomenology of gamma-ray bursts

Phenomenologically, classical gamma-ray bursts are brief (seconds to minutes) gamma-ray events which (a) originate from cosmological distances

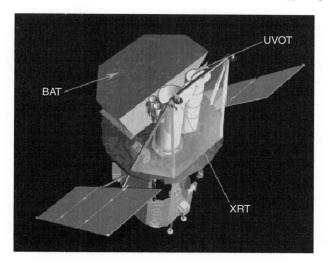

Figure 7.1   The Swift satellite showing the disposition of its gamma-ray Burst Alert Telescope (BAT), X-ray Telescope (XRT) and Ultraviolet/Optical Telescope (UVOT).
Source: NASA Swift team.

and (b) are one-time events (i.e., non-repeating[1]). A typical "light curve", representing the number of photons detected as a function of time, is shown in Fig. 7.2.

GRBs are detectable out to the furthest extragalactic distances. Although many of them have an identified host galaxy, the frequency of occurrence per galaxy is very small, ranging from 10^{-5} to 10^{-6} per year per galaxy, depending on burst and galaxy type, so none have been detected from nearby galaxies (let alone ours) in the last 30–40 years of observations. All indications are that they are connected to cataclysmic stellar events, which are likely to be connected either directly or after some delay to the formation of stellar mass black holes.

The gamma-ray spectrum, that is the number of photons per unit energy $N_E \equiv dN/dE$, is generally of the form of a broken power law (Fig. 7.3, upper panel). This is often also presented as the amount of energy in photons per decade of energy, given by $E^2 N_E$ (Fig. 7.3, lower panel), which is also a broken power law but with different slopes. The change of spectral slope occurs at a break energy E_b which, for the majority of observed bursts, is in the range of

---

[1] Unlike the soft gamma-repeaters discussed in Chapter 6, which are galactic events that repeat, related to neutron stars.

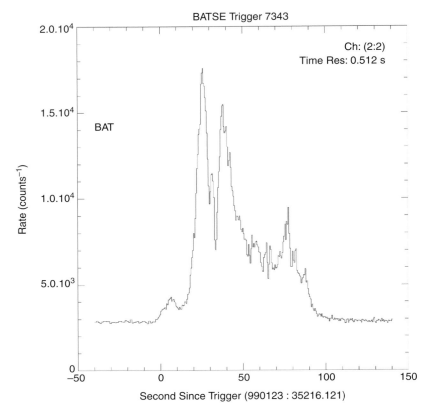

Figure 7.2   Gamma-ray light curve of GRB 990123 detected by the BATSE detector on Swift.

Source: NASA BATSE GRB team.

0.1–1 MeV. In a fraction of classical GRBs, this spectral break can look more rounded, as a quasi-black-body spectral hump, with power law extensions at higher and sometimes also lower energies. The high energy extensions in some cases extend well into the tens of GeV range, a feature which will be discussed further below and in Chapter 8.

There are at least two types of classical GRBs, which were initially distinguished by the duration t_γ of the gamma-ray emission (the gamma-ray "light curve") and by the energy of the spectral break E_b.

Long gamma-ray bursts. Long GRBs have MeV light curves (called "prompt" emission, to distinguish it from the subsequent longer lasting X-ray and softer afterglows) whose duration extends from about $t_\gamma \sim 2\,\mathrm{s}$ up to $t_\gamma \gtrsim 10^3\,\mathrm{s}$. These light curves are sometimes smooth, but often are highly and rapidly variable. An example of the gamma-ray light curve of a long GRB is shown in

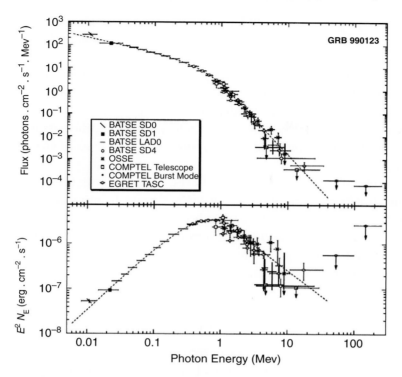

Figure 7.3   Gamma-ray spectrum of GRB 990123 from the BATSE detector, showing (A) a band-type broken power law component. The photon number spectrum $N_E \equiv dN/dE$ is in the upper panel, and the corresponding energy per decade spectrum $E^2 N_E$ is in the lower panel. (B) A possible second spectral component can be seen rising to the right.
Source: Reproduced by permission of the AAS.

Fig. 7.2. The spectrum of long GRBs is a broken power law with a typical break energy E_b in the range of 0.1–0.8 MeV, and some occur even lower, as low as a few tens of kiloelectronvolts. About two-thirds of all GRBs belong in the long category.

Short gamma-ray bursts. Short gamma-ray bursts are a clearly distinct class: their light curves are similar to that of Fig. 7.2 at MeV energies (which is the BATSE energy range), but compressed into a shorter interval of typically under 2 seconds, and sometimes as short as tens of milliseconds. The two duration classes are clearly separated: most long bursts are clearly longer than 10–20 s duration, and most short bursts are clearly shorter than 2 s (although some have a weaker and softer tail extending sometimes up to 100 s). The spectrum of short bursts is again typically of the broken power law shape, but with a harder break energy, typically $E_B \gtrsim 0.5 - 08\,\text{MeV}$, and sometimes as high as $3\text{--}4\,\text{MeV}$. For this reason, short bursts are sometimes called short hard bursts, and they comprise about one-third of all GRBs observed.

There is an obvious dearth of GRBs in the transition region around $t_\gamma \sim 2\,\text{s}$, although there appears to be a smaller number of objects which have intermediate durations and intermediate hardness spectra. These are referred to as a third group of intermediate GRBs. Their statistical distinctness is less strong than for the main long and short groups, on which most of the attention is focused.

The observed prompt gamma-ray fluxes, combined with the distances determined from detections of the host galaxies for which optical redshift distances are obtained, show that GRBs are the brightest explosions in the Universe since the Big Bang. If they were emitting isotropically, their energy output (in gamma-rays!) would on average amount to a solar rest-mass energy $M_\odot c^2 \sim 10^{54}\,\text{erg}$, give or take one order of magnitude,

$$E_{\gamma,iso} \sim 10^{53}\text{--}10^{55}\,\text{erg}, \tag{7.1}$$

all of which is emitted and gone in seconds. In fact, there is evidence that the emission is anisotropic (jet-like), with a typical jet opening angle θ_j of a few degrees, corresponding to a solid angle $\Omega_j \simeq \pi\theta_j^2$. This introduces for a double jet configuration an angular correction factor in the total energy emitted which, for long bursts, is on average $2\Omega_j/4\pi \sim 1/300\text{--}1/500$. Thus the actual average jet energy in gamma-rays is a less onerous but still stupendous

$$E_\gamma \sim 10^{51}\,\text{erg}, \tag{7.2}$$

emitted in a matter of seconds. This is to be compared with the (truly isotropic) kinetic energy content of a supernova explosion $E_{SN,kin} \sim 10^{51}\,\text{erg}$, only a very small fraction of which emerges as (mainly) optical photons over months to years. In fact, the gamma-ray energy output of a GRB in a few seconds is comparable to the energy output in optical light of the Sun over the lifetime of the Universe, $10^{10}\,\text{yr}$, or comparable to the luminous output of our entire galaxy over a century.

Thus, the big question in the early 1990s was, what can release such enormous amounts of energy in such a short time, mainly in the form of non-thermal gamma-rays? What is the "central engine" of such a monster, and how does it produce what we observe? It turns out that the second question, the how, is to a large degree more straightforward to answer than the what.

7.3   The GRB prompt radiation

Independently of the details of the central engine, and based only on the release of the above large energies on timescales of tens of seconds or less, the observed emission of gamma-rays and the afterglow must arise from a highly relativistically moving emission region. The enormous energy release in such short times in such compact regions produces a luminosity which exceeds more than a trillion times (by a factor $\gtrsim 10^{12}$) the critical Eddington luminosity $L_{Ed} \sim 10^{38}(M/M_\odot)\,\mathrm{erg\,s^{-1}}$ discussed in Section 5.1, above which radiation pressure overwhelms gravity. This huge outward pressure flings out the matter in the explosion region, which gets heated up into a fireball of electrons, positrons, gamma-rays and probably also some protons and neutrons, which will expand relativistically. For an amount of energy E injected into a fireball with an amount of rest-mass M in protons, the fireball is expected to accelerate until it reaches a terminal bulk Lorentz factor of

$$\Gamma = \left(E + Mc^2\right)/Mc^2 \simeq (E/Mc^2), \tag{7.3}$$

where the last equality holds for $E \gg Mc^2$. This just reflects the fact that fireballs endowed with more energy and carrying less mass reach higher terminal velocities (i.e., which approach c more closely).

The fact that one observes from GRBs photons with energies in excess of 30 GeV, for instance with the Fermi satellite, gives an estimate of how large the Lorentz factor Γ is. This is because photons produced in a plasma at rest would generally cross paths with each other at random angles of incidence, and photons whose energies ϵ_γ are above the electron rest-mass $m_e c^2 = 0.511\,\mathrm{MeV}$ by a certain amount can annihilate with other "target" photons whose energy ϵ_t is larger than $m_e c^2$ minus that same amount to give an electron–positron pair, a process symbolized as $\gamma\gamma \to e^+ e^-$. Thus, one would not have expected to see any photons with energies greater than 0.511 MeV, and yet one does see them. The only way out of this conundrum is if the plasma producing the photons is itself moving with a relativistic bulk velocity Γ. In this case the photons we observe have been collimated into a narrow cone of opening angle (in radians) of $\theta \sim 1/\Gamma$ along the jet axis, because the relativistic jet motion causes them, in the observer frame, to appear as if they are beamed along the jet direction. (The same occurs with parcels dropped from a moving airplane; even if they are dropped perpendicular to the plane's motion, in the frame of the ground observer the parcel is moving obliquely, having acquired a component of motion along the direction of the plane.) Thus, the beamed photons do not meet at large incidence angles but at shallow angles, and just as a grazing collision of two cars is less lethal than a head-on collision, the photon energy ϵ_t required for making an e^+e^- pair gets larger the shallower the collision angle, that is, the larger the jet bulk Lorentz factor.[2] Since the bulk of the "target" photons are those near the observed break energy of the broken power law spectrum of the GRB (see Fig. 7.3), $\epsilon_t \sim 1$ MeV, for observed high energy photons $\epsilon_\gamma \sim 30$ GeV one deduces that the jet must have a bulk Lorentz factor

$$\Gamma \gtrsim 200 \left[(\epsilon_\gamma/30\,\mathrm{GeV})(\epsilon_t/\mathrm{MeV})\right]^{1/2}. \tag{7.4}$$

More accurate Lorentz factor determinations are now being made, using Fermi observations and integrating over the full spectrum, which for instance have yielded in the case of GRB 080916C a value $\Gamma \sim 887 \pm 21$ [41].

These broken power law (Band type) spectral shapes are unlike the thermal black-body shape which would be expected from an opaque ("optically thick") fireball. In addition, a smoothly expanding fireball (in laminar flow) would convert during its expansion most of its internal thermal energy into bulk kinetic energy of the accelerated baryons and particles. That is, one would have an essentially cold, very fast moving fluid, and there would be very little thermal energy left in the particles to produce the very high energy photon flux observed. The way out of this paradox, which is the most widely held view, is that the bulk kinetic energy of the outflow is reconverted into thermal energy through shocks. Shocks have the property of using up bulk kinetic energy to convert it into random energy of particles. This random energy, usually called "thermal" energy (although in relativistic shocks it results in relativistic power law particle energy distributions), leads then to non-thermal photon emission looking like a power law, through the synchrotron or inverse Compton processes, if the shocks occur at radii beyond which the fireball has become "optically thin" (that is, the photon mean free path is larger than the size of the emission region, so the fireball is no longer opaque). This is the "fireball shock" scenario (e.g., [42]).

The gamma-ray light curves, as in Fig. 7.2, are often complicated and rapidly varying, sometimes on timescales as short as milliseconds. Such rapidly varying

---

[2] The threshold energy for the two-photon annihilation process depends on the relative angle of incidence θ of the two photons. This threshold condition is $\epsilon_\gamma \epsilon_t > (m_e c^2)^2/(1 - \cos\theta) \sim 4(m_e c^2)^2/\theta^2$. In a fireball moving with a relativistic bulk velocity v and a bulk Lorentz factor $\Gamma = (1 - [v/c])^{-1/2}$, special relativity indicates that particles, including photons, are only "aware" of other particles within their so-called light cone, inside a solid angle $\theta < 1/\Gamma$. That is, causality implies, from the above two-gamma annihilation threshold, that the bulk Lorentz factor satisfies eq. (7.4).

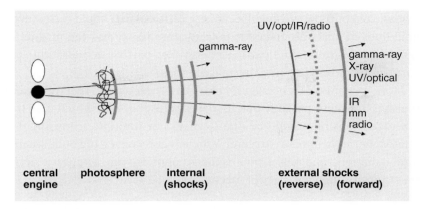

Figure 7.4 Schematic GRB jet and location of jet photosphere, dissipation (internal shock) and external shock radii.
Source: Bing Zhang.

light curves can be understood in terms of internal shocks in the outflow itself, caused by large internal velocity variations. These can arise if the outflow from the central engine is irregular, ejecting material whose velocity (or rather Lorentz factor) has large variations $\Delta\Gamma$ which are comparable to the Lorentz factor itself, $\Delta\Gamma \sim \Gamma$, occurring over a variability timescale t_v. The minimum variability timescale $t_v \gtrsim 10^{-3}$ s is likely to be intrinsic to the central engine, being comparable to the minimum orbital time of an accretion disk around the central BH of a few solar masses. In this internal shock scenario (Fig. 7.4), shells emitted from the central engine at an inner radius $r_0 \sim ct_v \sim 10^7$ cm at time intervals $t_v \gtrsim 10^{-3}$ s with Lorentz factors of order Γ which differ by $\Delta\Gamma \sim \Gamma$ would collide with each other at a larger radius given by

$$r_{dis} \sim 2ct_v\Gamma^2 \sim 6 \times 10^{12} \left(\frac{t_v}{10^{-3}\,\text{s}}\right) \left(\frac{\Gamma}{300}\right)^2 \text{cm}. \tag{7.5}$$

The resulting shocks would accelerate particles, including electrons and protons, to relativistic power law distributions, which would emit the non-thermal spectrum observed. There are issues to consider, such as the radiation efficiency, which is small unless particular conditions prevail, and work has been underway for some time on radiation mechanisms which result in higher radiation efficiencies.

One alternative prompt gamma-ray emission mechanism arises if the outflow is dominated by the magnetic fields in it (which could arise from the central engine, where dynamo mechanisms might occur). In this case, in the magnetized outflow the field lines of opposite polarity could reconnect. That is, field lines of opposite magnetic polarity annihilate, and the magnetic energy is converted into random particle energy (i.e., particle acceleration), and these particles would radiate non-thermal photons.

More recent evidence suggests that the characteristic spectral peak may be thermal in origin, possibly due to a jet photosphere, in which case the power law extensions may again be due to shocks or multiple scattering. Such sub-photospheric shocks are extremely efficient at reconverting bulk kinetic energy into radiation, and these photospheric photons can acquire a non-thermal spectrum through a variety of mechanisms.

7.4   GRB progenitors

The "large elephant" in the room is, of course, what are the actual objects which give rise to GRBs? In particular, since they occur in galaxies where previously there was no inkling that such an outburst would occur, one expects them to be related to some type of star which undergoes some cataclysmic event, and presumably there might be a difference between the stellar progenitors of long and short bursts, since their durations and spectral hardnesses are different enough. Historically, even before the CGRO satellite provided indications that GRBs are at cosmological distances, it had been suggested that the energetics required for a burst to arise from stellar objects could be achieved in the merger of a neutron star binary. Another event with roughly the right total energy, although apparently not the right timescales, is a supernova explosion. As it turns out, both of these prescient guesses are still the best guesses today, albeit with considerable elaborations.

7.4.1   Long gamma-ray bursts

In the observations of long GRB optical and X-ray spectra there is evidence for the presence of spectral line absorption of the continuum spectrum due to heavy elements in the local host galaxy medium, and the host galaxies are generally so-called star-forming galaxies, with active formation of young massive stars. The picture that has emerged as the most favored stellar progenitor of a long GRB is thus related to the most cataclysmic event expected from such young massive stars. Namely, that when the core of a massive star collapses, it leads to a black hole. Once the black hole has formed it will continue to accrete matter from the infalling stellar layers above it. Or if the star is somewhat less massive, the core collapse might lead first to a temporary neutron star or magnetar, which after some accretion also becomes a black hole. For a fast-rotating core, the infall of the overlaying layers will result in an accretion disk around the central compact object, with larger densities along the equatorial plane perpendicular to the centrifugally lightened rotation axis. For large accretion rates exceeding the Eddington rate, only a fraction of the accreted matter reaches the central object, the rest being ejected in a jet along the rotation axis by radiation pressure (or if magnetic fields play a dominant role, by magneto-centrifugal forces). This jet will continue to be fed as long as accretion continues, and provided it has enough momentum to overcome the weight of the overlying stellar material, it will emerge from the expanding outer layers of the star. The burst of gamma-rays would be associated with these emergent jets, whose properties allow us to infer (as discussed) that its motion is highly relativistic, with bulk Lorentz factors of several hundreds. According to numerical simulations, the accretion or jet-feeding can last tens of seconds, which is the right order of magnitude to explain the duration of long gamma-ray bursts.

7.4.2   *Long gamma-ray bursts and supernovae*

As a confirmation of such a core collapse origin as just discussed, in several cases where long bursts occurred in not too distant galaxies, a supernova of type Ic was detected at the same location, a few days after the GRB occurred.[3] The reason the supernova is detected optically only a few days later is that the supernova light is produced by the ejected stellar envelope, which is opaque to optical photons for several days, until the expansion has caused its density to drop sufficiently to allow their free escape.

Several such SN and long GRB coincidences have been discovered by Swift and HETE. Most notable has been the detection of the unusually long (~ 2000 s) soft burst GRB 060218, associated with the nearby ($z = 0.033$) SN 2006aj, a type Ic supernova. The exciting thing about this SN/GRB coincidence is that this was the first such event which was detected already in the first ~ 100 s also in X-rays and UV/optical. The early X-ray spectrum was initially dominated by a power law component, with an increasing black-body component which dominates after ~ 3000 s. This black-body component is thought to be due to the emergence of the SN shock from the stellar envelope as it broke through the optically thick wind of the progenitor.

---

[3] For the vast majority of long bursts this is not possible, because the distance is too large for a supernova to be detectable. Curiously, there are however a few examples where a long GRB was known to be near enough for a SN to be detectable, and yet none were detected down to $\sim 10^{-3}$ times the usual SN luminosity. It is unknown whether this means that in such cases an ejecta failed to be ejected, or whether some other explanation is required.

From a number of other SN/GRB coincidences, a trend that appears to be emerging is that, when a SN/GRB coincidence is observed, the GRB appears to be under-energetic compared to the average long GRB, while the SNIc is hyper-energetic (a "hypernova") compared to the average SNIc, or at any rate it has a faster expansion rate than a normal SNIc. This might suggest that when most of the energy goes into the GRB jet, there is little energy left for the envelope to produce a detectable SN-like emission, especially at the large distances of the majority of long GRBs; and conversely, when the GRB jet gets less energy, perhaps more energy is given to the ejected stellar envelope, producing an abnormally energetic SN-like emission coupled to a GRB jet which is only detectable if the distance is very close.

7.4.3   Short gamma-ray bursts

Short GRBs are believed to have a different type of progenitor than long GRBs. To begin with, short GRBs have never appeared in association with supernovae, and furthermore in a fraction of the cases where they have been believably localized, they appear to originate in elliptical galaxies known to have very little, if any, massive star formation going on. Compact binaries are in fact most abundant in old stellar population galaxies, such as ellipticals, although old stellar population components are present also in the halo of star-forming galaxies, such as spirals. This is compatible with the fact that short GRBs are indeed detected in comparable ratios in these two types of galaxies.

As suggested early on, another natural formation channel for stellar mass black holes is the merger of two neutron stars, each of which individually was the result of a supernova explosion of a progenitor star not massive enough to collapse into a black hole. While separately the two NS are stable, after they merge the mass of the central compact object exceeds the dynamical stability limit of a neutron star, and it becomes a black hole. Such mergers are expected when two neutron stars are in a binary system which formed either before or after the individual supernova events. Double neutron systems have been observed in a number of cases (e.g., binary pulsar systems), and gravitational wave radiation is inferred to lead to an observed gradual shrinking of their orbital separation (Chapter 6). It is expected theoretically that similar binary systems comprising a neutron star and a stellar mass black hole should exist, since some of the progenitor collapses can lead to stellar mass black holes. Another possibility for an indirect BH formation is if a white dwarf accretes so much mass from a binary companion that it collapses to a BH.

In all compact mergers leading to a BH, the mass of the latter comprises the larger fraction of the progenitor binary system. In both the NS–NS and the NS–BH mergers, rotation velocities approaching a fraction of the speed of light are achieved as the two objects approach within a few stellar radii of each other. This results in the disruption of the NS, and leads to a spinning BH with a rotating debris torus of nuclear density matter around it. The debris torus will be accreted by the black hole on a timescale estimated from numerical simulation to be on the order of seconds or less, which is the limiting duration of the hard radiation of short GRBs. The accretion, while it lasts, will also be super-Eddington, and will result in the ejection of a jet whose properties are somewhat different from those in long GRBs, since here there is no extended progenitor star to collimate the jet, but the outflow is inferred to be highly relativistic here as well, for the same reasons discussed above.

7.5   GRB afterglows

From the theoretical fireball jet model, it was expected that the prompt gamma-ray emission would be followed by a longer lasting and fading afterglow, whose photon energies would degrade sequentially from gamma through X-ray, optical and radio frequencies. The cause of this is that the relativistic outflow, while continuing to push a shock ahead of itself, is gradually slowed down by sweeping up increasing amounts of external matter (e.g., interstellar gas or the gas of the precursor wind of the progenitor star). Such afterglows were finally detected in 1997 by the Italian–Dutch Beppo-SAX satellite in X-rays, followed later by ground-based optical and radio detections.

Prompt localizations from space with the multi-wavelength Swift satellite have, since 2004, led to hundreds of follow-up observations of afterglows in optical, IR and radio, and since late 2008, the high energy Fermi satellite has also contributed additional detections of GRB at both MeV and GeV photon energies (see below).

The physics of the afterglow is relatively straightforward to understand. The fireball outflow moves outwards into an external environment which is tenuous, but is definitely not a vacuum. In the case of a long burst from a massive star's collapse, the progenitor massive star must have been sending out a stellar wind (all stars do, our Sun included) consisting of particles and magnetic fields, whose mass density is low and decreases with distance from the star. In the case of a short burst, if it arises from a merger of two compact degenerate objects, there is no, or precious little wind, and the fireball moves into the interstellar medium. The latter is a tenuous gas, whose density is approximately homogeneous on average. The outward moving fireball sweeps up this external medium, and compresses it as it goes through an external or forward shock, which advances into the unperturbed gas at the velocity of the ejecta. At the same time, the ejecta feels the ram pressure of the external material it is sweeping up, and this compression of the ejecta by the external matter sends a reverse shock moving into the ejecta. The ejecta and the swept-up gas of course continue to move forward together, but in the frame of reference of the interface between the ejecta and the external gas the reverse shock moves backwards.

The shock compresses the external gas of particle number density n cm^{-3} to a higher density $n\Gamma$, and heats it up to an energy per particle of $\Gamma m_p c^2$. When a substantial fraction of the initial fireball energy E has been used up in thus compressing and heating an amount of external gas which initially occupied a volume $V \sim r^3$, that is $E \sim r^3 n m_p c^2 \Gamma^2$, the fireball starts to decelerate. This defines a deceleration radius

$$r_{dec} \sim 5 \times 10^{16} \left(\frac{E}{10^{53}\,\text{erg}} \right)^{1/3} \left(\frac{n}{\text{cm}^{-3}} \right)^{-1/3} \left(\frac{\Gamma}{300} \right)^{-2/3} \text{cm}. \tag{7.6}$$

This external shock, or blast wave, and its associated reverse shock, accelerate the particles in both the shocked external gas and the shocked ejecta, which then radiate via synchrotron and inverse Compton radiation in a broad-band multi-wavelength spectrum [42]. After the fireball starts to decelerate at the radius (7.6), the expansion enters into a so-called self-similar phase, where all the relevant physical quantities evolve as a power law of the radius and of time. As the fireball continues to sweep up increasing amounts of matter, the blast wave bulk Lorentz factor progressively decreases. The fireball radiation, which is very highly boosted by large powers of the Doppler factor $\sim \Gamma$, becomes gradually less intense and more extended in time, resulting in an increasingly softer and longer lasting afterglow. Thus, the evolution of the external shock generates initially a prompt, hard spectrum, which subsequently evolves as a power law in time into an X-ray, then an optical and later a radio spectrum. The reverse shock, which is important only around the time when the initial deceleration occurs, results in a brief ultraviolet and optical flash. This generic afterglow model [40] has been confirmed in its main features, and remains the main workhorse for analyzing the data and comparing the observations to theory. There are, however, a number of interesting puzzles.

New insights on the burst and afterglow physics have been forthcoming from detailed X-ray light curves from Swift starting about 100 s after the trigger. Three of the features characterized by Swift have given rise in particular to much speculation. One of these is an initial very steep temporal decay of the X-ray afterglow, as seen in the left portion of the X-ray light curve in Fig. 7.5. The most widely considered explanation for this fast decay is that it is due to the

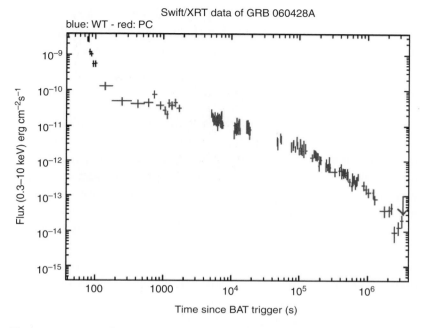

Figure 7.5   X-ray afterglow light curve of GRB 060428A detected with the XRT detector on Swift.

Source: NASA: Swift XRT team.

off-axis gamma-ray emission of the fireball, which not being directly aimed at us has a smaller Doppler boost and thus appears softer and arrives later than the on-axis prompt gamma-ray emission. This steep X-ray decay is often followed, in Swift observations, by a slower decay in time, seen as the flatter intermediate portion of the light curve in Fig. 7.5. A possible explanation is that a continued energy input into the fireball re-energizes the afterglow, so it does not decay so fast. There are other ways of achieving a similar result, for instance if the shock radiation becomes more efficient in time, which could be due to encountering more external material, or stronger magnetic fields, and for now it is unclear what the correct answer is. In addition, in many afterglows one or more steep X-ray flares appear superposed on the power law decay, typically between 100 s and sometimes as late as $> 10^5$ s, whose energy amounts to between a few and up to 50% of the total prompt emission. The rise and decay in time of these X-ray flares can be extremely steep, which is very hard to explain with any mechanism other than continued internal shocks or sudden dissipation, which implies a central engine activity lasting much longer than previously expected. Eventually, however, somewhere between an hour and a day, the self-similar power law decay expected from the simple forward shock fireball emission is regained (the right third of Fig. 7.5). In some bursts, this canonical behavior starts as early as tens of seconds, but the intermediate steep decay followed by flat decay behavior is encountered in about half the bursts.

7.6   Cosmological uses of GRBs

Long bursts are being increasingly found at redshift distances $z > 5$. For example GRB 050904, at $z = 6.29$, was located at a distance comparable to that of the most remote galaxies and quasars identified, being observed at a time when the Universe was less than 1/17th or 6% of its present age. This burst was extremely bright, with $E_{\gamma,iso} \gtrsim 10^{55}$ erg, and its X-ray intensity exceeded for a whole day that of the most distant X-ray quasar by a factor of up to 10^5. Even higher redshifts have been observed, for instance GRB 080913 at $z = 6.7$ and GRB 090423 at $z = 8.3$. The latter is the highest confirmed spectroscopic redshift of any object so far (July 2009), whether quasar, galaxy or GRB, it having occurred when the Universe was only 1/22nd of its present age.

The prospect of using such high redshift GRBs for determining cosmological parameters is tempting, but it is technically difficult due to problems in calibrating the GRB absolute luminosities as a yardstick, given their very large intrinsic variations. On the other hand, their extremely intense optical and X-ray radiation beams are excellent probes for absorption spectroscopic analyses of the intervening intergalactic medium, observed at redshifts when the Universe was being reionized by the first stars and galaxies. They can also provide a unique means of tracing star formation rates at very high redshifts, and have the potential to be excellent probes of the reionization era of the Universe, around redshifts 6–8, when the Universe emerged from its "dark ages" and luminous sources turned on.

7.7   Very high energy gamma-rays

Both leptonic (e.g., synchrotron and inverse Compton) as well as hadronic processes can lead to GeV and higher energy gamma-rays, both in the prompt and in the afterglow phases of GRB. The first moderate-significance detections of GRB in this energy range were obtained with the EGRET detector onboard the Compton Gamma Ray Observatory. The GeV emission detected by EGRET appeared in several bursts during the prompt phase (during which MeV photons are also observed), but in at least one case the GeV emission lasted for up to 1.5 hours after the trigger, during the afterglow phase. The Italian satellite AGILE was launched in 2007, with the goal of further exploring this interesting energy range, and it has been successfully measuring GRBs and AGNs.

The most powerful experiment in the GeV range is currently the Fermi Gamma-ray Space Telescope (see Chapter 8), launched in 2008, which has provided a new and powerful window into the very high energy behavior of GRBs. Roughly one GRB per week is detected with the Gamma-ray Burst Monitor (GBM, 8 keV–30 MeV), and roughly one a month is detected with the Large Area Telescope (LAT, 20 MeV–300 GeV). A number of bursts have been detected by the LAT at energies above 1 GeV, including several short bursts.

A notable long burst detected with the LAT was GRB 080916C, which had 14 events ranging from 1 GeV to 13.6 GeV, and over 200 events above 100 MeV [41]. This burst showed an interesting soft to hard to soft behavior, with a first peak in the MeV range only, but a second peak 3.5 s later with strong GeV emission. The MeV emission subsided after 55 s, but the GeV emission continued until 1400 s after the trigger. The spectra are mostly of the simple Band (broken power law) type, without additional spectral components hinting at either inverse Compton or hadronic effects (with rare exceptions). The spectra show an initial hardening and then a softening of the spectral break energy, while the high energy slope first steepens and then flattens. The lack of a clearly separate second spectral component in such cases suggests a single emission mechanism, possibly with varying emission parameters. A source with such a high density of GeV photons is extremely interesting, since in order for photons of this energy to avoid annihilation by interacting with other photons and converting into electron–positron pairs, the plasma jet where they originated must be moving at extremely relativistic speeds. In previous GRBs one had rougher estimates of how close the jet speed was to the speed of light, but with the large numbers of photons and excellent GeV spectrum of this burst it was possible to demonstrate that the jet had to be moving with Lorentz factors of $\Gamma \sim 880$, an extremely high value corresponding to a velocity which is roughly 0.999999 of the speed of light.

A most interesting result follows from the fact that in this and several other Fermi bursts the highest energy photons, those in the GeV band, arrived measurably later (seconds) than the lower energy MeV band photons which predominate at the outset of the burst. This is seen in the long burst light curves of GRB 080916C [41], and more importantly in the light curves of the short burst GRB 090510 [43] (Fig. 7.6), which impose stringent constraints on a broad class of quantum gravity theories. Quantum gravity is a theory which is as yet non-existent but which is the Holy Grail of 21st century physics, uniting gravity and quantum mechanics in a Theory of Everything (see Chapter 2). A generic prediction of many theories of quantum gravity is that they induce foam-like fluctuations in space-time which cause a relative delay between the propagation of higher and lower energy photons, a sort of vacuum dispersion effect.

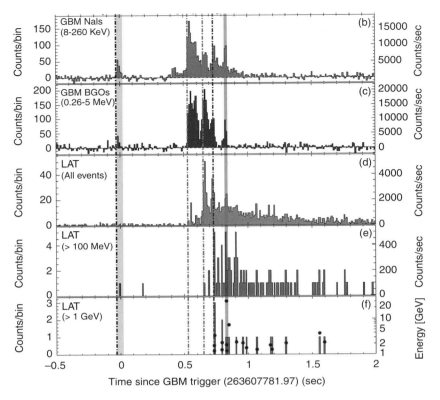

Figure 7.6   Gamma-ray light curves of the short GRB 090510 observed by the Fermi spacecraft at different energies [43].

Source: Reprinted by permission from Macmillan Publishers Ltd.

The magnitude of the delay depends on a fundamental quantity of physics, the quantum gravity energy scale E_{QG}, which is estimated to be around the Planck energy $E_{Pl} \simeq 1.3 \times 10^{19}$ GeV. For small enough delays, the delay Δt can be expressed as a series expansion in the small quantity E/E_{QG}, where E is the photon energy. The delay observed in GRB 080916C allowed the Fermi team to set an experimental lower limit to this scale of 1.5×10^{18} GeV for the first-order term in the expansion, just one order of magnitude below the Planck value [41].

Even more dramatic was the detection by Fermi of GRB 090510, the first short GRB to show clear emission in the LAT up to 31 GeV, a record at the time of measurement (see Fig. 7.6). This burst, in contrast to a number of previous ones, also showed for the first time a clear second spectral component, in addition to the usual Band-type simple broken power law [43]. However, it is as yet unclear whether this is of leptonic or hadronic origin. This burst also showed a time lag between the high and low energy emissions, allowing an even stricter limit on the quantum gravity energy scale. The experimental lower limit for the first-order term in this burst *exceeds* the Planck energy by a factor of 4, so the first-order term can be ruled out. This has serious implications for quantum gravity theories, which are beyond what can be discussed here [43].

The remarkable thing is that these unimaginably high energies around the Planck scale are completely out of reach of even the highest energy particle accelerators such as the LHC at CERN, which aims to probe up to 7000 GeV. Nonetheless, GRB observations of photons in the tens of GeV range allow us to set a robust experimental lower limit on this fundamental energy scale.

Both the leptonic and hadronic mechanisms emitting GeV photons should produce photons also in the TeV range, and intense searches at these energies continue to be made with ground-based air imaging Cherenkov telescopes (AICTs) such as HESS, VERITAS, MAGIC, and CANGAROO (see Chapter 8).

7.8   Non-photonic emission

A major question being investigated is whether this UHE gamma-ray emission is purely due to inverse Compton up-scattering of the MeV photons, or whether some part of this is associated with proton acceleration and hadronic cascades. In the case of hadronic cascades, a tell-tale signature would be TeV neutrino emission (see Chapter 11), since photo-meson interactions between the accelerated protons and photons in the source would also produce neutrinos at energies ranging from sub-TeV to EeV energies, where EeV $\equiv 10^{18}$ eV. Such neutrino signatures are being searched for with neutrino Cherenkov telescopes such as the cubic kilometer ICECUBE under the Antarctic ice, ANTARES, NESTOR and NEMO in the Mediterranean sea, and the planned cubic kilometer KM3NeT underwater telescope in the Mediterranean. Among experiments designed for the EeV range are balloon-borne radio detectors such as ANITA, which flies over Antarctica scanning an area of tens of thousands of square kilometers. The results of such experiments would provide critical information about fundamental neutrino interaction physics, as well as the particle acceleration mechanism, the nature of the sources and their environment.

The ultra-relativistic jets of GRB are thought to be capable of accelerating cosmic rays up to so-called Greisen–Zatsepin–Kuzmin (GZK) energies, $E_p \sim 10^{20}$ eV, above which interactions with the cosmic microwave background would impose a black-out beyond distances of 50 to 100 Mpc (see Chapter 10). It is estimated that GRB jets could result in a proton flux at Earth at these energies comparable to that observed with large extended air shower arrays such as the Pierre AUGER observatory. Also GRB-related hypernovae could be significant contributors to cosmic rays in the $10^{17}-10^{19}$ eV range, and may also produce TeV–PeV energy neutrinos.

Another important type of non-photonic emission from GRBs which is being searched for is gravitational waves (see Chapter 9). Gravitational waves require an event which has a time-varying mass quadrupole moment. That is, it requires a mass distribution that does not keep spherical symmetry, and is not just simply oscillating back and forth along an axis, but is varying along two independent directions, such as a binary system of two masses orbiting around each other. Gravitational waves are expected at some level from all stellar collapses, if the collapse has a chaotic behavior, such as blobs forming with a radial plus a random velocity component. Thus, all core collapse supernova remnants are (weak) sources of gravitational waves. Long GRBs, being core collapses of massive stars requiring extreme core rotation rates, may be stronger sources of gravitational waves than normal core collapse supernovae. However, the most promising sources are short GRBs, if these are indeed compact (NS–NS or NS–BH) mergers. Such signals are being actively sought with the LIGO and VIRGO gravitational wave observatories.

7.9   Wider impact of GRB multi-channel studies

The critical importance of a multi-wavelength approach for understanding high energy phenomena generally and GRBs in particular is worth stressing. It was not until other wavelength ranges were brought to bear that astrophysicists were able to solve the distance scale and figured out the progenitors. In turn, the multi-wavelength attack on GRBs has been crucially enabled, at several stages, by the capability of satellites to deliver rapid, accurate positions to ground-based observers. This is an important lesson from multi-wavelength electromagnetic (photon) observations. It is likely that opening up new non-photon observing channels, such as neutrinos or gravitational waves, will lead to even greater progress. For instance, if gravitational wave bursts are observed for short GRBs seen in gamma-rays, this would be a "smoking gun" proof of the compact binary merger hypothesis. Or if a TeV neutrino burst is observed from a GRB seen also in gamma-rays, this would confirm the theoretical expectation that protons are accelerated in GRB jets – and could contribute to the ultra-high energy cosmic ray flux.

Conversely, the feedback from GRB investigations to multi-wavelength studies and discoveries of other astrophysical phenomena is remarkable. Studies of GRBs have revealed tremendously interesting phenomena in the X-ray/optical/radio sky that were completely unknown before GRB observers and observatories came along that were capable of telling astronomers when and where to look. This has included naked-eye optical flashes from a GRB at a cosmological distance of $z = 0.9$; making spatially resolved observations (images) of an expanding relativistic radio source at $z = 0.17$; discovering the existence of supernovae with relativistic ejecta; finally discovering the shock breakout from a core collapse supernova as it occurs; etc. Thus, GRB studies illustrate vividly the wider impact of high energy astronomy as a discipline – without the high energy "trigger" it is doubtful we would know of these phenomena, even today.[4]

---

[4] I am indebted to my colleague Derek Fox for reminding me of these points.

8

GeV and TeV gamma-rays

8.1 Importance of the GeV–TeV range

The GeV–TeV gamma-ray range holds a strategically important role in astrophysics, by providing the first high quality surveys of most classes of very high and ultra-high energy sources, including sufficiently large numbers of objects in each class to be able to start doing statistical classifications of their properties. The number of photons collected for individual sources in this energy range extends in some cases into the tens of thousands, leading in many cases to quite high signal-to-noise ratios.

The GeV–TeV photon emission provides not only important information about the photon emission mechanisms and the source physical properties, but also clues for the importance of the corresponding very high energy (TeV and up) neutrinos and even higher energy cosmic rays which may be emitted from such sources [44]. In addition to the discrete astrophysical sources, instruments in this energy range also provide information about the diffuse gamma-ray emission, such as that associated with cosmic rays interacting with the gas in the plane of our galaxy, the diffuse emission from our galactic center, and the extragalactic emission component, all of which could yield information or constraints about possible dark matter annihilation processes, in addition to the astrophysical processes and the sources involved.

8.2 Galactic Gev–TeV sources

Pulsars. Among the earliest galactic sources discovered at GeV energies are the rotation-powered pulsars. As already discussed in Chapter 6, the large dipole magnetic fields ($B \sim 10^{12}$ gauss) of pulsars – combined with their

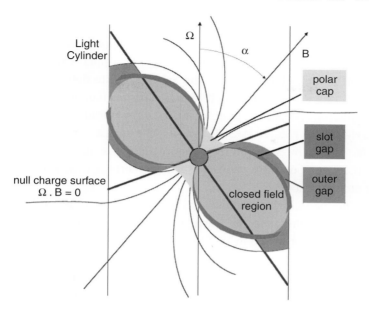

Figure 8.1   Pulsar polar cap and outer gaps where accelerated electrons can produce gamma-rays.
Source: NASA.

rotation – lead to the formation of extremely high electric field regions across low density regions in their magnetosphere called "gaps" (see Fig. 8.1). Depending on the orientation of the rotation axis (Ω) and the magnetic dipole axis (*B*), the acceleration and radiation could be produced dominantly in the so-called polar cap gaps located near the stellar surface at the magnetic polar caps. Alternatively, the dominant pulsar radiation may arise further out, at the so-called outer gaps located close to the cylindrical surface called the "light cylinder", where the magnetic fields swept around by the rotation approach the speed of light, shown by the two outer lines parallel to the rotation-axis in the figure.

The gamma-ray radiation is expected to be pulsed (Fig. 8.1), whether it arises from the polar gaps or from the outer gaps, as long as the rotation axis does not coincide with the magnetic axis, for similar geometrical reasons as the radio radiation is seen to be pulsed. The radio, optical, and X-ray radiation is thought to be produced by accelerated electrons as they move in curved paths, by so-called curvature radiation and synchrotron radiation. The GeV emission is generally thought to be due to inverse Compton scattering of softer photons to GeV energies by the energetic electrons. Six pulsars were discovered by the

EGRET experiment on the CGRO to show regular GeV pulsations. However, it was not clear from the off-set between the radio, optical, X-ray, and gamma-ray pulses whether the geometry corresponded to a polar cap or to an outer gap emission model. More recently, the Fermi satellite detected new pulsars showing GeV pulsations, with much higher statistics (i.e., with a much larger number of photons per pulse). In at least two of the pulsars observed by Fermi, PSR J0205+6449 and PSR J2021+3651 [45, 46], the emission pattern is fan-beam shaped, that is the emission is directed more along the magnetic equator than along the magnetic polar direction. This is in agreement with what is expected from an outer gap model.

On the other hand, so far no pulsar has been detected to pulse at TeV energies at the spin period, which may be due to a lack of electrons energetic enough to scatter the softer photons up to TeV. In the binary pulsar PSR B1259-63/SS2883 [47], there is a regular TeV periodicity but this is at the binary period, and it is thought that a shock occurs where the magnetized pulsar wind collides with the companion stellar wind, accelerating electrons to the higher energies needed to upscatter photons to TeV energies.

Supernova remnants and pulsar wind nebulae. Other sources of both TeV and GeV emission in the galaxy are shell-type supernova remnants (SNR) and pulsar wind nebulae (PWN). The former are mature supernova remnants which are no longer influenced by the central compact remnant, for example the SNR

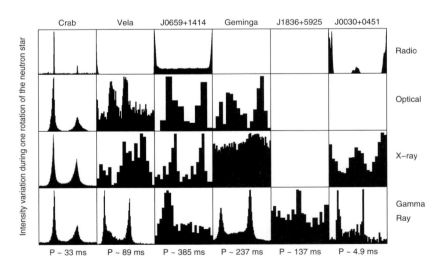

Figure 8.2 Light curves as a function of spin phase for seven pulsars, with gamma-ray data from Fermi, and the corresponding X-ray, optical and radio pulses. *Source*: D. Parent and the Fermi LAT pulsar group.

remnant Cas A. The latter are younger supernova remnants into which a central pulsar's wind injects an appreciable energy. The first source ever to be detected in the VHE gamma-ray range was the Crab nebula (Fig. 4.6), a well-known example of a pulsar wind nebula. Its detection by the Whipple Observatory in 1989, by Trevor Weekes and collaborators, marked the realistic launch of TeV gamma-ray astronomy [48].

SNRs are perhaps the earliest recognized "high energy" astrophysical sources. In the late 1940s and early 1950s they were observed to show non-thermal (power-law like) and polarized radio and optical spectra, which indicated a synchrotron mechanism from relativistic electrons with a power-law energy distribution. Since they are left over from SN explosions where highly energetic nuclear reactions play a dominant role, they were long suspected to have other high energy photon and possibly hadronic signatures as well. Later on SNRs were detected in X-rays ($\sim$ 0.1–10 keV), most recently with the Chandra and XMM spacecraft; low energy gamma-rays (0.1–10 GeV), for example with the EGRET experiment on the CGRO satellite; and now with the LAT experiment on the Fermi spacecraft. They have also been detected in VHE gamma-rays (0.1–10 TeV) with ground-based AICTs such as HESS, VERITAS, MAGIC, CANGAROO, and others [49].

Morphologically, SNRs are of two main types: shell type and filled-in (plerion) type. The shell type are generally younger remnants, and constitute the bulk of the GeV–TeV gamma-ray detected ones (over a dozen now). In shell remnants, such as SNR 1006 shown in Fig. 6.3 in a Chandra image, the harder X-rays are non-thermal, presumably due to synchrotron radiation, and concentrated in thin filaments in the outer limbs. The same electrons can scatter softer photons, such as abundant cosmic microwave photons up to GeV or even TeV energies. A number of shell SNRs have in fact been detected at GeV and/or TeV energies. An example is the SNR G347.3-0.5, seen in a HESS TeV image with a superposed X-ray image in Fig. 8.3. The TeV emission, however, may also arise from the decay $\pi^0 \rightarrow 2\gamma$ of neutral pions into high energy photons, where the pions are produced by nucleon–nucleon collisions of relativistic protons accelerated in the same shocks as the electrons, colliding against thermal protons in the remnant. This hadronic interpretation of the TeV spectrum competes with the leptonic inverse Compton scattering hypothesis (called leptonic because it involves electrons, as opposed to nuclei). It is still unclear which of these two interpretations are correct, although it appears clear that at least in a number of SNRs the hadronic interpretation is disfavored. However, from other observations and theoretical arguments discussed in Chapter 10, SNRs are the likeliest sources of the observed cosmic rays (protons) up to energies $E_p \lesssim 10^{15}$ eV, so there ought to be electromagnetic decay products (i.e., TeV photons) of hadronic

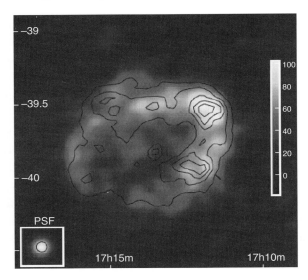

Figure 8.3 Supernova remnant G347.3-0.5 image in TeV gamma-rays (shaded, from HESS) and in X-rays (black contours, from ASCA). *Source*: HESS team.

processes associated with these cosmic rays. The problem is that it is difficult at present to distinguish them from the (also unavoidable) leptonic contribution.

X-ray binaries. Some TeV and GeV sources are binary X-ray sources, including such well-known X-ray binaries as Cygnus X-1, and objects such as LS 5039 and LS I+61 303. These are generally classified as either microquasars (see Section 6.9; i.e., sources with a relativistic jet resulting in shocks), or else as binaries with a stellar wind-driven shock, leading to particle acceleration and gamma-ray emission.

8.3   Extragalactic sources

Active galactic nuclei. As discussed in Chapter 5, AGNs are broadly classified as radio quiet and radio loud, the radio quiets having weak or no jets and being located mainly in spiral galaxies, and the radio louds generally having prominent jets and being often hosted in large elliptical galaxies. Sub-types of radio-loud AGNs include radiogalaxies and radio-loud quasars, which besides detectable jets generally also have significant nuclear emission (see Fig. 5.4).

Among the radio-loud AGNs the highest energy radiation is observed from *blazars*, whose emission is dominated by their jets which are aligned close to the observer line of sight. Two blazar sub-types discussed in Chapter 5 are the BL Lacertae (BL Lacs, for short), which exhibit no broad emission lines, and the flat spectrum radio quasars (FSRQ), which do show broad emission lines ("flat" referring to the shape of the radio spectrum). Most of the observed luminosity of blazars is in VHE gamma-rays, extending to the GeV–TeV range. Besides the non-thermal radiation component associated with the nucleus or its jet, both these sources and quasars in general also have less luminous continuum radiation (UV or X-ray) associated with an accretion disk, and also emission line components associated with clouds further out. In blazars, however, which have jets with bulk Lorentz factors of order $\Gamma \sim 10$–30 which are observed almost head-on, the resulting very high Doppler boost results in an extremely intense jet non-thermal continuum brightness, which almost completely overwhelms the line and thermal continuum from the host galaxy, and also results in very high photon energies. Nonetheless, the disk continuum photons and the cloud line photons, even if unobserved, play an important role as seed photons for inverse Compton upscattering and/or photo-hadronic processes such as $p + \gamma \to \pi^0 \to 2\gamma$, leading to VHE photons in both cases.

The shape of the gamma-ray spectra of blazars is influenced both by the jet Lorentz factor and by the angle at which the jet lies to the line of sight. Many blazars, as well as a few other AGNs such as M87 or Cen A, which are detected only up to the GeV range (e.g., by the EGRET, AGILE and Fermi spacecraft), have jets which are not too closely aligned to the line of sight, $\theta \sim 3$–$10°$, and have Lorentz factors $\Gamma \sim 2$–5. This gives a less extreme Doppler boost, and also allows in some cases the observation of superluminal expansion velocities, which require intermediate angles. Those blazars which are detected in the TeV range by ground-based AICTs have much smaller angles to the line of sight $\theta \lesssim 1$–$2°$, and larger inferred Lorentz factors $\Gamma \sim 10$–30. As mentioned, the blazar spectra are generally double-humped broadband spectra extending from radio to gamma-rays, with a first peak in the optical to X-rays, and a second peak in gamma-rays from MeV to GeV or TeV. An example is shown in Fig. 8.4. Depending on the hardness of these peaks they are classified as "low" peak blazars (LBL) and "high" peak blazars (HBL). There are also intermediate peak objects, not surprisingly called IBLs. Some of the most notable TeV blazars include Mrk 421, Mrk 501, PKS 2155, and FSRQ such as 3C 279.

The two-humped spectra are suggestive of a combination of a synchrotron plus an inverse Compton spectrum (i.e., a mechanism involving only electrons and thus referred to as the leptonic model of blazars[1]). The lower optical to X-ray spectral hump is clearly non-thermal in origin, and the inferred electron synchrotron radiation must arise from the inner jet portions at radii in the range

---

[1] A similar two-humped spectrum can also be expected in SNRs and in GRB from the same mechanisms, as discussed in Chapters 6 and 7.

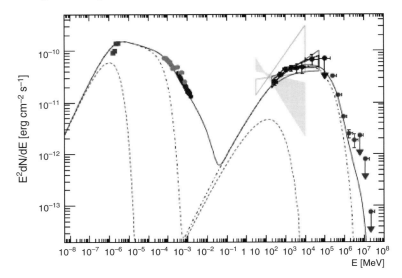

Figure 8.4 Broad-band spectrum of the blazar PKS 2155 observed by HESS and Fermi [50].

10^{14}–10^{16} cm. Such radii are needed to explain the variability times $\Delta t_{var} \lesssim r/c\Gamma^2 \sim 3 \times 10^3 r_{15}\Gamma_1^{-2}$ s $\sim$ hours with modest jet Lorentz factors $\Gamma \sim 10\Gamma_1$. On the other hand, the gamma-rays must come from radii $\gtrsim 10^{16}$–10^{17} cm in order to be above the "pair photosphere" where the $\gamma\gamma \to e^+e^-$ process mean free-path (Section 8.4) is of order unity, otherwise pairs are produced and the higher energy photons (which are observed) would have disappeared. However, the radio emission of the jets must come from $\gtrsim$ pc scales, in order to avoid self-absorption.

Leptonic blazar models. In leptonic models of blazars, the higher spectral peak is ascribed to inverse Compton scattering radiation, which may involve different components. For instance, the IC process may involve electrons upscattering the synchrotron photons which they themselves previously produced (synchrotron-self Compton, or SSC), or it may be that they scatter other, externally originated photons (synchrotron-external Compton, or EIC). A generic leptonic gamma-ray blazar model is shown schematically in Fig. 8.5. This shows the various possible sources of seed photons (e.g., UV photons from the disk, the wind or from BLR clouds) located at the right distance $\lesssim 10^{17}$ cm for upscattering by relativistic electrons in the jet at $r \sim 10^{16}$–10^{17} cm. Relativistic electrons at the appropriate height are expected to be produced by internal shocks, similarly to the GRB internal mechanism, due to irregular modulation of the relativistic outflow with varying Γ. Such leptonic models can be calculated in a fair amount of detail.

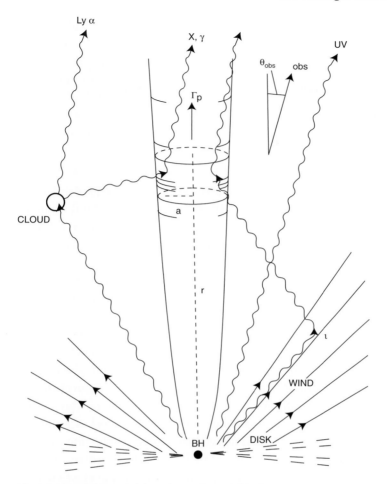

Figure 8.5 Schematic blazar leptonic model [51].
Source: Reproduced by permission of the AAS.

Hadronic blazar models. Alternatively, the higher spectral peak could be due to hadronic effects, which, as in the case of SNRs and GRBs, can produce high energy GeV–TeV photons from hadronic cascades initiated by protons accelerated in the same shocks as the electrons. This is the generic hadronic model of AGN, although again, there are different versions (for instance, another version involves proton synchrotron for producing TeV photons). In the hadronic cascades, protons interact with photon targets leading to charged pions, which decay into (among other things) secondary charged muons, positrons and electrons, which then radiate synchrotron radiation in the magnetic field and inverse Compton scatter, leading to new e^+e^- pairs, etc. Such cascades are discussed in more detail in Chapter 10. The reason the hadronic interpretation is attractive is because charged hadronic cosmic rays are observed at Earth in the $10^{17}-10^{20}$ eV range (Chapter 10) and these are almost certainly extragalactic, and besides GRBs the other main candidate astrophysical sources for such cosmic rays are AGNs. As also in SNRs and GRBs, the shocks which accelerate electrons must also be able to accelerate protons and ions. The ambivalence between the hadronic cascade and the leptonic model interpretations for the higher hump remains so far unresolved.

Gamma-ray bursts. Classical GRBs are definitely extragalactic, and in fact detectable out to the highest spectroscopically confirmed redshifts, the current record as of 2009 being held by GRB 090423 at $z = 8.3$. More than 3000 GRBs have been detected at MeV energies, most of which are at a redshift $z \geq 1$, and it is estimated that at least 10% are at $z \gtrsim 5$. Of these, four were also detected at GeV energies by the EGRET experiments in CGRO, and as of November 2009 scores have been detected with the LAT detector onboard the Fermi satellite, at energies above 100 MeV, with at least seven at energies above 1 GeV, as discussed in Chapter 7.

Tantalizingly, we expect the spectrum of GRBs to extend into TeV photon energies, especially the afterglow, since in the external shock the internal target photon density is lower and $\gamma\gamma$ absorption is not likely within the source (e.g., [42]). On the other hand, significant $\gamma\gamma \rightarrow e^+e^-$ absorption is expected in the intergalactic medium, unless the source is at low redshifts (say less than 100 Mpc; see Section 8.4), and precious few GRBs are expected at these distances. Indeed, as of 2009, searches had not yielded any GRB detections at TeV energies with a high enough confidence level (meaning, in technical terms, greater than 5σ significance detections). However, some GRBs do occur close enough to avoid significant absorption, and detecting them at TeV remains an actively pursued goal.[2]

8.4   Detectability of GeV–TeV sources

GeV and TeV gamma-rays are observed from a number of different astrophysical sources, though not always both in the same objects. The absence of either GeV or TeV in any particular object is not necessarily due to a preferential emission by the source in one or the other energy band, but is sometimes related

---

[2] An additional problem is that TeV air Cherenkov telescopes are optical detectors and can only operate on moonless nights, so the active duty cycle is only 10% of the time, and bursts are both brief and sporadic, at most only a few a year being expected within the absorption-free distance.

to the distance of the sources and the available observational techniques. Very high energy (GeV and above) gamma-rays of energy $\epsilon = \hbar\omega$ are subject to absorption as they travel in the intergalactic medium by interacting with ambient "soft" photons of some lower energy $\epsilon_s = \hbar\omega_s < \epsilon$, producing electron–positron pairs. This process is symbolized as

$$\epsilon + \epsilon_s \rightarrow e^+ + e^-. \tag{8.1}$$

The VHE photon of energy ϵ and the soft photon of energy ϵ_s annihilate each other and become an electron–positron pair. This process was already discussed in Chapter 7 in connection with gamma-ray burst sources, where the target ϵ_s were within the source itself. Here we consider the same process once the photons ϵ have escaped the source and are making their way through intergalactic space, where a different set of ϵ_s await them. The reason the annihilation results in a pair of particles (e^+ and e^-) is, first, that electric charge must be conserved (zero total charge before and after) and second, that there were two photons with different momenta to begin with (and to conserve momentum one needs at the end two other particles). Finally, the total energy must also be conserved, taking into account that the initial photons have energies but no rest-mass, whereas the two final electrons have both a kinetic energy and a rest-mass energy of $m_e c^2$ each. It can be shown that for the two photons to annihilate and create $e^{\pm}$ conserving both momentum and energy, the photon energies must satisfy a threshold condition, which (if we assume that the photons are more or less isotropic, i.e., ignoring angle effects) can be roughly expressed as saying that the geometric mean of the energies of the two photon energies must exceed two electron rest-masses:

$$(\epsilon.\epsilon_s)^{1/2} \gtrsim 2m_e c^2 \simeq 1\,\text{MeV}. \tag{8.2}$$

This relation shows that two photons of energy $\epsilon \sim \epsilon_s \gtrsim 1\,\text{MeV}$ can produce an $e^{\pm}$ pair, but more interestingly, photons whose energy ϵ is much larger than an MeV can pair-produce even in the presence of target photons ϵ_s much less energetic than an MeV, provided their product satisfies the threshold condition (8.2). Thus, photons of $\epsilon \sim \text{GeV} = 10^3\,\text{MeV}$ can already pair-produce with soft photons of energies $\epsilon_s \gtrsim 10^{-3}\,\text{MeV} = \text{keV}$, and photons of $\epsilon = \text{TeV} = 10^6\,\text{MeV}$ pair-produce with photons $\epsilon_s > 1\,\text{eV}$ (i.e., optical photons).

The opacity of the intergalactic medium to high energy photons is determined by the above threshold condition, and of course, by the probability that the "projectile" photon ϵ encounters a target photon ϵ_s along its route to the observer. The latter is clearly going to be proportional to the target photon density n_s (cm^{-3}), since there must be enough of them for the probability of a photon–photon interaction to be non-negligible over the path length of the VHE photon. The more targets there are per unit volume, n_s, the higher the chance that in each cubic centimeter an interaction leading to annihilation occurs. Even if the photon does not interact in the first cubic centimeter containing n_{γ_s} such targets, the longer it travels the higher the probability that eventually it will interact with a target. The probability of an interaction leading to annihilation is thus proportional to the target density n_s, to the path length traversed r (cm), and to a quantity called the interaction cross-section for the $\gamma\gamma \rightarrow e^{\pm}$ process, $\sigma_{\gamma\gamma}$ (cm^2). This cross-section is a measure of how effectively a single target photon can "block" an incoming projectile photon to produce an e^+e^- pair, and it is non-zero only for incident and target photons which satisfy the $\gamma\gamma$ threshold condition (8.2). The larger the target density and the cross-section, the shorter is the path traversed by the VHE photon before it annihilates. One can define a "mean free path" λ for photons of energy ϵ, $\lambda(\epsilon)$, which is the average path length that photons ϵ can go in a medium with a density $n_s(\epsilon_s)$ of target photons of energy ϵ_s before producing an $e^{\pm}$ pair,

$$\lambda(\epsilon) = \frac{1}{n_s(\epsilon_s).\sigma_{\gamma\gamma}}. \tag{8.3}$$

This photon mean free path (m.f.p.) is plotted in Fig. 8.6 for VHE photons propagating in the intergalactic medium, considering the three most abundant types of target photons: the cosmic microwave background radiation (MBR), the infrared and optical (IR/O) background photons produced by stars, and radio photons from radio galaxies. In these soft target photon fields, VHE photons of energy $\epsilon \equiv E$ (eV) have the mean free paths (Mpc) plotted in the ordinate of Fig. 8.6.

As can be seen in Fig. 8.6, the mean free path of 100 GeV (10^{11} eV) photons is at least 1000 Mpc, depending on various assumptions (labeled a, b, c) about the infrared/optical (IR/O) poorly known target photon background density at various redshifts (particularly in the IR, where detectors are not as advanced as in the optical range). This target photon background is mainly due to stars, and the rate of star formation at high redshifts is also only fragmentarily known. The lower limit of 1000 Mpc is however quite conservative. The cosmic microwave background (MBR) has photons of mean energy $\sim 10^{-3}$ eV which mainly annihilate incident photons of $\epsilon \sim 10^{15}$ eV, whose m.f.p. is ~ 10 kpc. Higher energy incident photons would have increasing m.f.p., and such photons are discussed in Chapter 10.

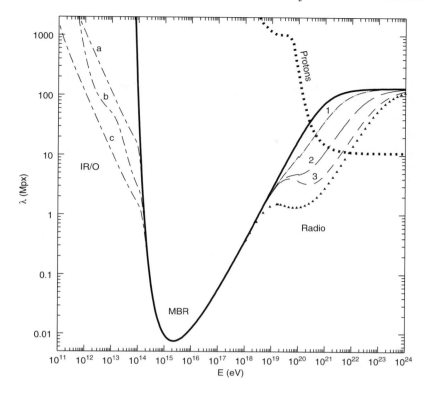

Figure 8.6 Mean free path to $\gamma\gamma \to e^{\pm}$ for photons of energy $\epsilon = E$ against IR/O, MBR and radio photons in intergalactic space [52].
Source: Reproduced by permission of the AAS.

The lesson from this figure is that the Universe is essentially transparent to 10 GeV photons, whose m.f.p. $\lambda > 1000$ Mpc corresponds for a standard cosmological model to a redshift $z > 0.35$. This distance is comparable to or larger than the "effective" visible radius of the Universe (the Hubble horizon), given by the speed of light times the Hubble time $t_H \sim 10^{10}$ yr, that is $r_H \sim ct_H \sim 4000$ Mpc. On the other hand, the Universe is fairly opaque to TeV photons, whose m.f.p. may be as small as $\lambda \sim 100$ Mpc, corresponding to a redshift of $z = 0.024$ for a standard cosmological model. One expects very many extragalactic sources within a m.f.p. of 10 GeV photons, but rather few within a m.f.p. of 1 TeV photons. Thus, TeV sources can be measured (if not too weak) from inside our own galaxy, but even very luminous extragalactic TeV sources cannot be detected unless they are at relatively small distances. Of course, the IR background is poorly known, so the above statement must be treated with caution. However,

GeV sources, if bright enough, can be detected both from our own galaxy and also from up to the most distant objects in the Universe.

One technical caveat here is that the Earth's atmosphere is rather opaque to GeV photons, so these have to be observed from satellites above the atmosphere, limiting the weight and size of the detectors. TeV or higher energy photons, on the other hand, can be measured from the ground, allowing much larger detectors to be built (see Section 8.5). Another caveat is that different source emission spectra, combined with possible energy dependence (e.g., $\gamma\gamma$ absorption) within the source itself, can create entirely different spectra as seen at the observer.

8.5   GeV and TeV detection techniques

Photons of GeV energy cannot reach the ground because they convert into e^+e^- pairs high in the atmosphere, and the pairs dissipate their energy without producing any signal at ground level. Thus, GeV photons are detected from balloons or from space. There have been several satellites at GeV energies, the two most recent ones being the AGILE mission, launched by the Italian Space Agency, and the Fermi satellite, operated by an international collaboration and launched by NASA.

Fermi has two main experiments on board. One is the Large Area Telescope, sensitive to photons between 20 MeV and 300 GeV, with a 2.5-radian field of view and an angular resolution of arc-minutes. This consists of stacks of high energy pair-conversion counters (see Fig. 8.7). A high energy photon interacts with an electron or a nucleon in one of the upper modules creating an e^+, e^- pair along the direction of the photon but at a slight angle to each other, depending on the photon energy, and these are tracked as they travel down in the stacks. At the bottom of the stacks there is a calorimeter, a device where the pairs that got through are stopped and their energy is thermalized. At the eight corners of the spacecraft are eight lower energy detectors constituting the GBM, which is sensitive in the 10 keV–30 MeV range, comparable to the range of the BATSE detector on CGRO, and with comparable sensitivity and 4π field of view. Fermi does not have the capability to slew fast towards a location where GBM detects a transient source outside the LAT field of view (uploading and executing a re-pointing takes hours), but the joint probability of either its own GBM or the Swift BAT observing the same source as LAT is substantial, and such simultaneous observations of GRBs are accumulating.

Photons of TeV energy have the benefit that they produce signals which are detectable at ground level. Of course they interact with electrons or nucleons at the top of the atmosphere leading to an e^+e^- pair, which initiates a pair

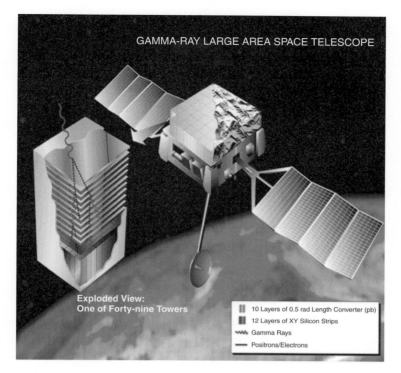

GAMMA-RAY LARGE AREA SPACE TELESCOPE

Exploded View:
One of Forty-nine Towers

	10 Layers of 0.5 rad Length Converter (pb)
	12 Layers of XY Silicon Strips
	Gamma Rays
	Positrons/Electrons

Figure 8.7 The Fermi Gamma Ray Observatory (previously called GLAST) houses the LAT, sensitive at GeV energies, and the GBM, sensitive at MeV energies. *Source:* NASA.

cascade. The positron annihilates with another atmospheric electron producing another lower energy gamma-ray, and the electron as well as the positron can also inverse Compton scatter optical photons to create lower energy gamma-rays, which promptly again pair-produce, leading to a shower of relativistic $e^{\pm}$ pairs propagating downwards (see Fig. 8.8).

These relativistic $e^{\pm}$ pairs travel in the Earth's atmosphere at a speed which slightly exceeds the phase velocity at which light can propagate in the atmosphere. This is not possible in vacuum, but a finite density medium such as air (or water) has a refractive index which results in light propagating at speeds ever so slightly below the speed of light in vacuum. As a result, the relativistic $e^{\pm}$ can radiate optical/UV photons which, amazingly, travel slower than the electrons do, in a phenomenon called Cherenkov radiation. (This is similar to airplanes which fly supersonically and can outrun their own sound, which arrives in a sonic boom after the plane has passed.) The $e^{\pm}$ eventually exhaust themselves and the shower snuffs itself out in the upper atmosphere, but not before it has radiated lots of optical Cherenkov photons, which reach the ground.

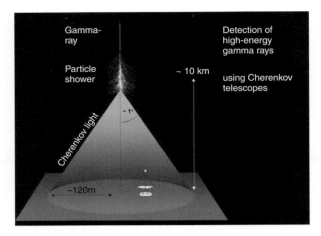

Figure 8.8   air Cherenkov telescope principle.
Source: W. Hoffman and the HESS group.

The cascades start at a height of about 10 km, and the cone of Cherenkov light is about a degree wide for TeV photons (it depends on the speed of the $e^{\pm}$, hence on photon energy), so the footprint of a single TeV photon above the atmosphere is a circle of about 100 m diameter on the ground. An optical telescope conveniently placed detects the Cherenkov optical light produced by the TeV photons. Of course, cosmic rays (primary charged nucleons and electrons incident on the atmosphere) also produce cascades which Cherenkov radiate, but the characteristic properties of cosmic-ray cascades are different from those of photon-induced cascades, as is the distribution of the resulting photons in the sky, and both of these properties are used to distinguish them. Large single-dish telescopes can also provide good spectra of sources, and low resolution images.

Higher resolution imaging is greatly aided by locating multiple telescopes within a distance smaller than the Cherenkov light cone of a single dish, which provides a stereoscopic view of the sources. Thus, if the source is spatially extended and not too distant (e.g., supernova remnants in our galaxy), one obtains spatially resolved images in TeV light. An example is the supernova remnant G347.3-0.5, obtained with the HESS four-telescope imaging array (Fig. 8.3). HESS, located in Namibia (Fig. 8.9), has recently been upgraded, adding a fifth, larger telescope at the center of the four shown in the figure. Another major air Cherenkov array is VERITAS, consisting of four telescopes located in Arizona, USA, shown in Fig. 8.10. The CANGAROO array is another four-telescope

Figure 8.9 The HESS Cherenkov telescope array, showing the first four dishes.
Source: The HESS team.

Figure 8.10 The VERITAS air imaging Cherenkov array in Arizona, status as of
November 2009.
Source: S. Criswell, Whipple Observatory.

array located in Australia, while the large (17 m diameter) MAGIC telescope in
La Palma, the Canaries, has recently been upgraded with a second dish to a
two-mirror stereoscopic AICT configuration.

9
Gravitational waves

9.1 Ripples in space-time

The gravitational force field, as discussed in Chapter 2, is described in General Relativity as a distortion of space-time caused by the masses in it, which results in any small test mass in this space-time moving along the curvature of the space-time. If the position of the large source mass (or masses) which dominate a certain region of space-time is varying, the space-time structure readjusts itself to reflect the changed positions of the source masses, after a delay caused by the fact that the information about this change of position of the source masses cannot be communicated faster than the speed of light. That is, the space-time at some location r away from the source mass which has moved can respond to this change only after a time $t = r/c$. This traveling information about changes in the space-time structure is the basis of the phenomenon of gravitational waves, which can be thought of as ripples in the texture of space-time that travel at the speed of light.

One can visualize this also in a simpler quasi-Newtonian picture, provided one accepts the relativistic principle that information travels at most at the speed of light. Imagine two equal masses M in a circular orbit of radius d around each other, in a plane parallel to the line of sight to the observer, with the center of mass of the orbit (the mid-point of the line separating the two) being a fixed point in space at a distance D from the observer (see Fig. 9.1). Assume that the period of rotation of the orbit is P minutes, and suppose that at some instant $t = 0$ the two masses, as they appear in the sky at position A, have their largest projected apparent separation. This will also mean that they are both at the same distance D from us, the observer at the point O in the figure, one a bit to one side and the other a bit to the other side of the line of sight to the

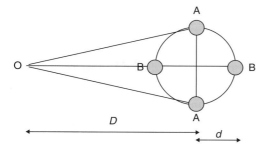

Figure 9.1   Varying gravitational field from a binary.

central imaginary point separating them. Since they are at the same distance from us, the force will be just twice the force from each. Now after a time $P/4$ the two masses will have rotated to position B, where one is behind the other. We see only one, but we feel the gravitational force of both, since that does not get blocked. However, the nearer one has a stronger force than either of the two at the previous instant in position A, since it is closer, while the force from the farther one is weaker than that from the nearer one, and also weaker than either of the two at A, since it is farther away. However, the sum of the forces from the nearer and the farther at B is still a bit larger than the combined force of the same two masses at the earlier position A. This difference in force means that test masses in our lab at O, following Newton's second law that acceleration equals force divided by mass, will be accelerated differently by these distant masses M. Now after a time $P/2$, the two masses will appear in the sky again to be at a maximum apparent separation at position C, but being indistinguishable, this position is equivalent to A and the force will be the same as at A. Thus the acceleration felt by the test masses in our lab will vary periodically with a period $P/2$ (half of the orbital period!). This regularly varying acceleration causes the test masses in our lab to react regularly with the same period $P/2$, in response to the varying gravitational force. The response motions in our lab, however, occur after a time delay of D/c relative to the time at which the changes of orbital position occurred, since the two masses are at an average distance D away, and any information concerning the changes of position travels with a velocity given by the speed of light c. The above explanation, however, is not the whole picture: it just shows that the gravitational field at the observer position O varies on a certain timescale, but it has not yet told us how it varies. In order to do this, we must discuss the symmetries of the source mass distribution.

One condition required to produce gravitational waves is that the varying source mass distribution must have a so-called mass quadrupole moment. That is, it cannot just simply be a mass distribution such as, say, a star which expands and contracts uniformly in a "breathing" mode (i.e., a monopole). Neither can it be a rotationally symmetric disk that pancakes in and out along its smaller dimension into a narrower or wider height disk (a dipole). The sources that can emit gravitational waves must undergo time-dependent oscillations of a more complex nature, the simplest of which have a quadrupole moment. At the simplest level this means that the "source" is squashing in along one axis, and stretching out along a different axis, which is one way of looking at the pair of masses rotating in orbit around each other in Fig. 9.1. This quadrupole nature of gravitational waves is in contrast to the basic mode for electromagnetic radiation, which is dipole. The difference is that gravitation, unlike electromagnetism, does not have charges of opposite sign, mass having only a single sign.

Thus, gravitational waves represent fluctuations of the curvature of space-time, and curvature does not influence a single test particle at a particular position but rather the relative positions of test particles at two different positions. The fact that curvature affects the relative position of two reference points can be illustrated by means of the analogy of two skiers gliding down the same downward-curving slope. The skier who is further ahead on the steeper part of the downward slope moves faster and gets increasingly separated from the other skier who is moving in the same path but is further behind in the less steep portion of the slope.

This response is somewhat similar to that of the tides in the oceans which occur in response to the attraction of the Moon, which causes a stretching of the water mass along the Moon–Earth direction (let's call this direction z), and squeezes it inwards in the directions perpendicular to the Moon–Earth direction (let's call these directions x and y). On the other hand, the response of the test masses to a gravitational wave is to undergo oscillatory motions, moving in along one axis x, and moving out along a perpendicular axis y; half a period later the roles of x and y reverse, with the test masses moving out along x and in along y. This is illustrated in Fig. 9.2. In contrast to the Earth–Moon (A) longitudinal tidal stretching and squeezing, in the gravitational wave case the test mass stretching and squeezing are entirely transverse (B and C), both being in the x, y plane perpendicular to the source–detector direction z.

The detectors of gravitational waves exploit this property (see Section 9.5). The response motions of the test masses in our lab are somewhat similar to the motion of buoys bobbing up and down as a regular train of ocean swells passes under them. The regularly varying gravitational field is a passing gravitational

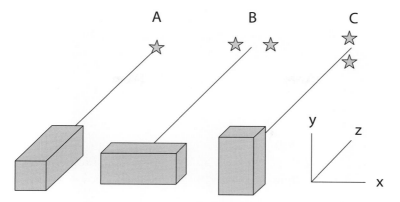

Figure 9.2   As opposed to the effect of a single star (A), the effect of the varying mass quadrupole of a binary (in positions B and C) causes an alternate stretching and squeezing along axes perpendicular to the line joining the binary and the lab test mass.

wave. In the language of General Relativity, the motion of the distant masses induces changes in the space-time around them, which propagate outwards in all directions at the speed of light, just like water waves in a pond when we drop a stone into it, and these changes in space-time are felt in our lab, in response to which our test masses execute regular motions as they "fall" along the regularly varying space-time curvature perturbations.

9.2   Astrophysical sources of gravitational waves

From the previous section it is apparent that a promising gravitational wave (GW) source would be two masses in orbit around each other, such as binary stars, binary galaxies, or as we shall discuss later, binaries consisting of compact objects such as neutron stars or black holes. It is useful to see first some of the general properties of such binary GW sources (Fig. 9.3).

The GW luminosity L would be expected to depend on how strong the gravitational force is between the two masses. Thus, it ought to depend on how large the masses are, and since the force is proportional to the product of the masses, for equal mass stars the luminosity is proportional to the square of the masses, M^2. Also, it should depend on the separation ℓ between the masses, being larger for binaries whose separation is larger (i.e., larger for binaries for which the varying gravitational field is larger). Furthermore, one would expect that binary systems whose motion is fast (i.e., those where the orbital period P is shorter) should radiate more efficiently than those which are varying slowly (in the same way that a fan which is rotating faster makes a stronger breeze).

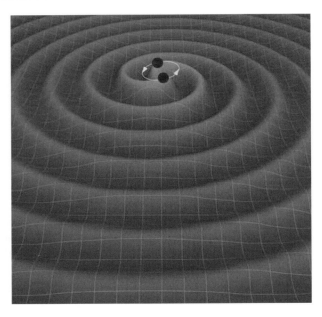

Figure 9.3   Gravitational wave schematic.
Source: NASA.

General Relativity tells us that the luminosity is proportional to the product of
the masses and the ratio of the fourth power of the orbital size (ℓ^4) and the sixth
power of the period (P^6). Additionally, Kepler's law, which was first applied to
planetary orbits in our solar system, shows that the binary system's total mass
$M = M_1 + M_2$, orbital size ℓ, and period P are related so that the binary system's
mass is proportional to the ratio of the cube of the orbital size to the square of
the orbital period. Putting it all together, we have

$$L_{GW} \sim \frac{GM^2 \ell^4}{P^6} \propto \left(\frac{M}{\ell}\right)^5 , \tag{9.1}$$

where G is Newton's gravitation constant.

This GW luminosity means that gravitational energy is being carried away
from the source, in this case the binary system. Gravitational waves carry
energy, since they cause test masses far away to bob up and down, or left and
right, which requires some energy being delivered to them by the wave. This
can be understood by considering that the gravitational waves lead to a force
that stretches or squeezes the space between nearby test masses in our lab. If
we made the masses beads on a rod, then the stretching or squeezing of space
would lead the beads to move on the rod (assume the rod has a fixed length,
which does not change). As the beads move, friction turns their velocity on

the rod into heat; so, gravitational waves can transfer energy from the binary system to heat the rod in our laboratory. The change in separation of the test masses is proportional to the product of the separation before the waves pass and the amplitude of the wave, which is called the strain h. It can be shown that the strain of a gravitational wave at a distance r from a GW source where masses M are in orbits of size ℓ is

$$h \propto \frac{G^2 M^2}{\ell r}. \tag{9.2}$$

Thus, the amplitude of the response varies inversely proportionally to the distance r of the source.

The gravitational energy loss of a binary means that the source is left with less energy. However, all gravitationally bound systems, such as a binary, have a negative total energy. Their total energy is the sum of their kinetic energy of motion, which is positive, and their gravitational potential energy, which is negative, and the gravitational dominates the kinetic. If the kinetic energy were the dominant (e.g., if they were rotating too fast, they would be flung apart, like two children holding hands who spin too fast, lose their grasp and fly apart). The reason binaries are binaries (i.e., they stay together) is because the negative gravitational energy (the grip of the hands) dominates over the positive kinetic energy of motion. Now, if a system has negative total energy, and it loses some more energy due to GW emission, it must be left with less energy (i.e., its negative total energy must have become even more negative). The gravitational potential energy is $E_g = -GM^2/\ell$, so since M has not changed, it means that the separation ℓ must decrease, making E_g more negative. At the same time the period shortens, following Kepler's law, so the positive kinetic energy of motion increases too, but in such a manner that it does not offset the increasingly negative gravitational energy, and the net result is that the binary becomes more tightly gravitationally bound as it loses energy through emitting GWs. Thus, the binary orbit shrinks, as a result of the gravitational wave energy loss.

9.3   Stellar binary GW sources

About half of all stars in the galaxy are in binary systems, so in principle it would seem that there should be many stellar GW sources. However, the GW energy loss formula [eq. (9.1)] shows that most binary stellar systems should be very weak GW sources. This is because for the majority of stellar binaries the orbital separations are too large, or equivalently, the orbital periods are too long for L_{GW} to be important. That is, if we look at the value of L_{GW} implied by the masses and separation values of most stellar binaries, and we let them lose energy by GW emission over a timescale comparable to the age of the Universe, the Hubble time $\sim 10^{10}$ years, the energy of the binary would have changed too little to make any significant difference in the separation ℓ due to this process. This is true for most binaries consisting of "normal" stars such as the Sun, and even more so for giant stars of radii much larger than the Sun: even if the two stars were so close as to graze each other, their GW energy loss would be small.

However, the situation is different for compact binaries – that is, binaries made up of compact stars, such as white dwarfs, neutron stars or stellar black holes. White dwarfs (WD) have typical radii $R_{WD} \sim 10^9$ cm, which is about 1% the radius of the Sun, while being $\sim 10^3$ times the radius of a neutron star or a stellar mass black hole. Normal stars with masses less than $\sim 8 M_\odot$ end up being white dwarfs, while larger ones end up being neutron stars or black holes. However, the galaxy's "mass function", the number of stars formed per unit mass, is a steeply decreasing power law of the mass, there being many more low mass than high mass stars in a typical galaxy. Thus, one expects many more white dwarfs accumulating in the Universe as a result of stellar evolution than either neutron stars or black holes. This means large numbers of white dwarf binaries, which is good news. The orbital separations ℓ cannot be smaller than the WD radii $\sim 10^9$ cm, which means their orbital periods P are tens of minutes or longer. The frequency of the GW would be

$$\nu_{WD} = (2/P) \lesssim \left(\frac{GM}{\ell^3}\right)^{1/2} \sim 10^{-3} \left(\frac{\ell}{10^{11}\,\text{cm}}\right)^{3/2}\,\text{Hz}, \tag{9.3}$$

where we took a separation $\ell = 10^{11}$ cm and masses $0.5 \lesssim M_{WD} \lesssim 1.4 M_\odot$. This is the millihertz range at which the LISA GW detectors would be sensitive. The corresponding GW strains from eq. (9.2) are such that there should be plenty of them which are sufficiently nearby to be within reach of LISA.

Neutron star (NS) binaries would have comparable masses to WD binaries, but since they have 10^3 times smaller radii they can in principle reach much smaller orbital separations, $\ell \gtrsim 10^6$ cm, and periods P down to milliseconds, with GW wave frequencies ranging up to the kilohertz range. The strains, from eq. (9.2), would be larger than those for white dwarfs, but the waves would have much higher frequencies. This is the range in which the LIGO (Fig. 9.4) and VIRGO (Fig. 9.5) gravitational wave detectors are sensitive. Known neutron star binaries have been detected thanks to the fact that one or both of the members of the binaries are radio pulsars, and this allows us to determine, through the regularly varying Doppler shift of the pulsations, the orbital parameters such as the separation, the ellipticity and inclination of the orbit, etc., as well as the masses of the components. In some cases the binary may consist of a pulsar

Figure 9.4   LIGO Gravitational Wave Observatory near Hanford, Washington.
Source: Photo courtesy of the LIGO Laboratory.

and a black hole, and stellar population synthesis calculations indicate that one should also expect compact binaries made up of two stellar mass black holes. The latter, however, have not been detected electromagnetically so far, since there is no matter to accrete which could radiate, and furthermore black holes, as the saying goes, "have no hair", meaning here that they have no magnetic fields of their own which would radiate like a pulsar.[1]

All compact binaries, especially ones with smaller separations, are expected to have substantial GW energy losses, meaning that their orbits should shrink in a detectable manner. The binary loses both energy and angular momentum as it emits gravitational waves, and eventually the orbit should shrink until the stars come in contact. In the case of NS–NS, NS–BH or BH–BH binaries, at the instant of merger the velocity of rotation is $v \sim c$ and the frequency of the GW is twice the rotation frequency, so $\nu_{GW} = 2\nu_{rot} = \omega_{rot}/\pi = (1/\pi)(GM_{tot}/\ell^3)^{1/2}$, assuming $M_{tot} = 2M$, or

$$\nu_{NS} \sim \left(\frac{GM_{tot}}{\ell^3}\right)^{1/2} \simeq \left(\frac{c^3}{71GM}\right) \sim 10^3 \left(\frac{3M_\odot}{M}\right) \text{Hz}, \tag{9.4}$$

where for the minimum separation we took roughly two Schwarzschild radii, $\ell \sim 2R_S = 4GM/c^2$. As the binary approaches the merger, the frequency ramps up gradually in a characteristic "chirp" signal, which should be a tell-tale signal encoding characteristics of the binary. For such a binary at a distance $r \sim 300\,\text{Mpc} \sim 10^{27}$ cm we have, from eq. (9.2), a dimensionless strain $h \sim 10^{-21}$.

---

[1] Whatever magnetic field they had at the time of their initial collapse through the light horizon would have been swallowed into the horizon in a few dynamical (free-fall) times. Hence John Wheeler's dictum that black holes have no hair.

Figure 9.5 VIRGO Gravitational Wave Observatory view near Pisa, Italy.
Source: The VIRGO team.

This is near the sensitivity capabilities of the advanced versions of the LIGO and VIRGO laser interferometric gravitational wave antennas, which are optimized for the 100 Hz frequency range (9.4) (i.e., the frequency range of the GW waves emitted as the NS or BH binaries are approaching merger).

The rate at which mergers can be expected is estimated both from radio observations of binary pulsars and from population synthesis models of the evolution of binary stellar systems. Observationally, in a number of cases the shrinking of the orbit of a compact binary has been measured at separations well before the merger, the first case being the Hulse–Taylor double pulsar [53]. For this pulsar system, the orbital period decay rate and the orbital velocity are measured to be

$$\Delta P \simeq 77 \, \mu s/\text{year}, \quad (v_{orb}/c) \sim 0.15\%, \tag{9.5}$$

a small but measurable effect, which confirms the theoretical decay predicted from general relativistic gravitational wave energy losses. There are several other detected radio pulsar binaries for which such gravitational wave energy losses are measurable via radio observations, which allow a determination of the secular change of their orbital motion. Thus, there is a fair observational handle on how many such binaries exist, and how many per year would be expected to merge within a certain distance from us. The most reliable current estimates are that the advanced LIGO and VIRGO systems should observe NS–NS or NS–BH inspirals at a rate of between two per month and one per day [54], and these compact binary mergers are prime targets for these GW observatories.

Aside from purely GW detections of such binaries, simultaneous observations in other channels may also play an important role. For instance, as discussed in Chapter 7, the most likely model for short gamma-ray bursts is that they are neutron star binary mergers ejecting a relativistic jet which emits gamma-rays. In this case, one expects that the simultaneous detection of a short gamma-ray burst and a simultaneous burst of gravitational waves should greatly enhance the detectability of the latter.

9.4 Galaxies as gravitational wave sources

Galaxies pass by each other occasionally, especially if they are members of a galaxy group or cluster. In our own "local" group of galaxies there are some two-score plus members, of which the Milky Way and Andromeda (M31) are by far the largest members, which orbit around each other. The masses are of course large, but so is the separation (0.7 Mpc), so the GW emission expected from such a binary is negligible.

In rich clusters of galaxies, with a thousand or more galaxies of masses comparable to or larger than our own, the spatial density of massive galaxies is large enough that binaries are common and some have smaller separations. In fact, in many clusters there occur close enough encounters or collisions between galaxies where significant perturbation of the individual galaxy structure occurs, ripping out gas and stars from each other, which are left as debris trails in the wake of the collision. The gas churning in the violently varying gravitational field of the two galaxies leads to shocks which result in greatly enhanced diffuse electromagnetic emission in X-rays and optical, as well as in bursts of enhanced star formation. The gravitational wave emission from such close passages or from the binaries thus formed is also rather weak, as far as the galaxy-scale dark, stellar and gas mass distributions involved.

However, most galaxies appear to be endowed with massive black holes at their core, as discussed in Chapter 5. Our own Milky Way has an MBH of $\sim 3 \times 10^6 M_\odot$, other normal galaxies have larger MBHs, and AGNs require even larger MBHs of $10^8 - 10^9 M_\odot$ to explain their very large non-thermal multi-waveband luminosities and also to explain their highly energetic non-thermal jets. When either normal or AGN-type galaxies collide or interact, their MBHs represent the largest and most concentrated gravitational potential region in them. If the two galaxies merge, the violent relaxation of the combined gravitational potential leads to a readjustment of the mass distribution where the heaviest objects sink to the new combined center, and the two MBHs find themselves in orbit around each other in the newly formed galactic core. The dynamical friction in the chaotically varying gravitational field of the rest of the dark matter and stars, lasting for several dynamical times following the merger, leads to a fairly tightly bound MBH binary. Accretion of gas onto, and disruption of nearby stars by, the MBHs leads to their being detectable as point sources of electromagnetic radiation in images of the merged galaxy, confirming the above binary formation picture, dozens of them having been identified. From such observations, which provide values for the separations and from mass estimates based on accretion luminosities, we can conclude that many of these binary MBHs must merge within a Hubble time, leading to powerful gravitational wave emission.

The typical frequency of the GWs emitted at the instant of merger of two MBHs of $M = 10^6 M_\odot$ is

$$\nu_{MBH} \sim \left(\frac{GM_{tot}}{\ell^3} \right)^{1/2} \simeq \left(\frac{c^3}{71GM} \right) \sim 3 \times 10^{-3} \left(\frac{10^6 M_\odot}{M} \right) \text{Hz}, \tag{9.6}$$

for a separation approximately equal to two Schwarzschild radii, $\ell \sim 2R_S \sim 4GM/c^2 = 6 \times 10^{11}(M/10^6 M_\odot)$ cm. This is the GW chirp frequency as the binary builds up towards the final merger. This millihertz range is the frequency range for which the LISA space-based GW detector has been optimized. It is estimated that several such mergers may be within reach of LISA at all times, in various stages of approaching their final merger [55]. See Fig. 9.6.

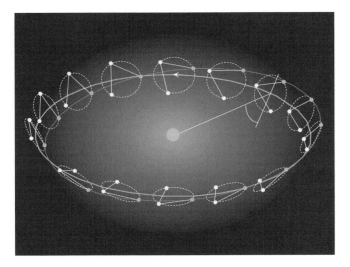

Figure 9.6 The planned LISA three-detector array will face the Sun, at an angle of 60° to the plane of Earth's orbit, revolving with Earth around the Sun.
Source: NASA, ESA.

9.5   Gravitational wave detectors

Gravitational wave detectors are built around the principle of measuring the changes of the separation x between test masses induced by the strains h of eq. (9.2), the relative change being $(\Delta x/x) \sim h \sim 10^{-21}$ for typical estimates. These are obviously extremely small changes, which are very difficult to measure, but this has not deterred people from attacking this problem.

The earliest detectors were cylindrical bars of metal with piezo-electrical sensors at the end which could measure the minute changes of electrical current induced by the very small changes of elongation of the two ends of the bar. The main difficulty is to eliminate the natural background of extraneous mechanical vibrations, such as trucks passing near the lab, footsteps, seismic noise, etc., which can be partly alleviated by suspending the bar in a vacuum. Then there is the other problem of the natural thermal motions of the molecules of the bar, which are reduced by cooling it to extremely low temperatures. This type of detector continues being developed today, with many technical improvements.

A second generation of GW detectors is built around the principle of detecting the small periodic changes induced by the passage of the GW in the separation between pairs of reflecting mirrors. A laser beam is bounced between the mirrors many times, and the changes in separation are measured by the technique of Michelson interferometry, measuring the interference fringes of the laser light with itself (see Fig. 9.7). The laser on the bottom left shoots a beam which goes through a beam splitter (a semi-reflecting mirror) in the center, half the

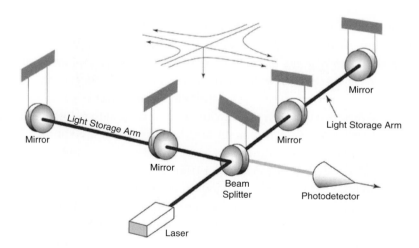

Figure 9.7   GW laser interferometric detection scheme.
Source: Photo courtesy of the LIGO Laboratory.

beam going on and being reflected by the mirror on the upper right, while the other half beam which had been deflected to the left by the splitter is reflected by the mirror on the upper left. The beams undergo repeated reflections between outer mirrors and inner mirrors (test masses), until the two beams are reunited at the splitter and go on to a photo-detector on the lower right. If the lengths of the two perpendicular arms of the interferometer have changed relative to each other, this will show up in the reunited laser beam, since the two waves will interfere with each other, showing dark fringes where they interfere destructively and bright fringes where they reinforce each other. The oscillation of the mirrors is measured by the frequency and displacement of the fringes. The two sets of mirrors on the upper right and upper left act as the test masses being moved by the gravitational wave.

The LIGO and VIRGO detectors are built on the above principle, and were designed to measure GWs in the compact binary merger frequency range of hundreds of hertz, with best sensitivity around 150 Hz. The length of the perpendicular arms is 3–4 km, LIGO consisting of two such interferometers, one at Hanford, in the state of Washington (Fig. 9.4) and another similar detector in Livingston, Louisiana, whose arms are at 45° orientation relative to those of the Hanford detector, in order to increase sensitivity to both the possible polarization modes of the GWs. The VIRGO detector (Fig. 9.5) is located near Pisa, Italy, and has similar characteristics and sensitivity as the LIGO detectors. The three sets of detectors are operated together as a giant intercontinental interferometer array. The current sensitivity is approaching $h \sim 10^{-20}$, and the planned sensitivity of the advanced versions of these detectors, after planned upgrades, is expected to reach the critical $h \lesssim 10^{-21}$ values.

The LISA detector is similarly based on the Michelson interferometer principle, but in order to be sensitive to the millihertz (10^{-3} Hz) frequencies expected from massive black hole mergers and galactic white dwarf binaries it must have much more widely separated arms. This is because the wavelength $\lambda = c/\nu \sim 3 \times 10^{10}/10^{-3} \sim 3 \times 10^{13}$ cm is comparable to an astronomical unit, the distance between the Sun and the Earth, and to get good interference patterns one needs interferometer arm lengths which are comparable to this distance. Thus, LISA is planned as a space interferometer, and consists of three independent satellites arranged in a triangle, with reflecting mirrors at each station to bounce laser beams among themselves (Fig. 9.6).

The frequency range and the planned sensitivities of LISA and LIGO, with VIRGO having similar values to LIGO, are shown in Fig. 9.8.

Another proposed way of detecting gravitational waves is through the effect they have on the pulsed radio signals of pulsars. The pulsar signals are very regular (Chapter 6), and the small ripples in space-time induced by gravitational

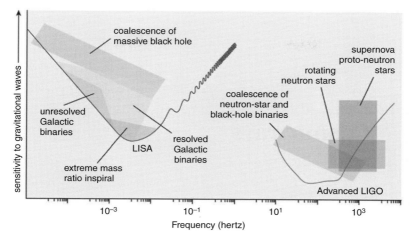

Figure 9.8 LISA and LIGO estimated sensitivities.
Source: LISA project (ESA, NASA, AEI). Used with the permission of Dr T. Prince LISA Mission Scientist, Caltech, USA.

waves can show up as disturbances in the regular pulsar signals. The sensitivity of pulsar timing arrays has increased dramatically over the last several years. The North American community has organized itself under the banner "NANOGrav", for North American Nanohertz Observatory (for) Gravitational-waves. In the European Union scientists have launched a similar project called LEAP (Large European Array for Pulsars), and in Australia there is a corresponding project called PPTA (Parkes Pulsar Timing Array). All three of these organizations are fully cooperating and sharing timing data under the International Pulsar Timing Array (IPTA) banner [56]. If Advanced LIGO remains on its present schedule (operations beginning in 2014) and the pulsar timing arrays continue to improve as in the past several years, there is a good chance that the first gravitational wave detection might be with these pulsar timing arrays. These signals could be either from individual super-massive ($\sim 10^9$ solar mass) black hole binaries, or a stochastic background arising from the superposition of many super-massive black hole binaries.

10

Cosmic rays

10.1 Particles from Heaven

Cosmic rays are energetic particles that reach us from outer space, arriving from all directions. They are generally electrically charged particles, such as protons, heavy nuclei, electrons and positrons, but more broadly one includes among them also electrically neutral particles such as neutrons and neutrinos from outer space. If one subtracts those that arrive from the Sun, the rest arrive essentially isotropically, constituting a uniform background of cosmic-ray radiation, made up of particles with a finite mass. In addition to these, there is also a separate photon background, which includes the cosmic microwave background, the diffuse starlight optical-infrared background, and X-ray and gamma-ray backgrounds, all of which are also essentially isotropic, after subtraction of individual resolved sources.

A major difference between the cosmic-ray background and the photon background is that photons are massless and electrically neutral, so they travel essentially in straight lines from their sources, making it (at least at some wavelengths) easier to identify where they ultimately came from. The vast majority of cosmic rays, however, are electrically charged, and this makes it far harder to discern where they came from. This is because the interstellar and intergalactic space is woven through by random magnetic fields, and the Earth's atmosphere is permeated by an ordered magnetic field, so that as a result of propagating through these magnetic fields the cosmic ray path has little to do with the direction of whatever source they originated from [57]. This is because the deflection of the cosmic ray at any point along their path in a magnetic field is perpendicular to both the instantaneous cosmic-ray velocity vector direction

$\vec{v}$ and to $\vec{B}$, the local magnetic field direction.[1] This is a generic property of how magnetic fields act on moving electric charges: the force exerted by a magnetic field is always sideways to the direction of motion of the particle. As a result, a relativistic particle of mass m, electric charge Ze, velocity (normalized to the speed of light) $\beta = v/c$, Lorentz factor $\gamma = 1/\sqrt{1-\beta^2} \geq 1$ moves in a corkscrew motion centered along the direction of the magnetic field $\vec{B}$. Its longitudinal motion along the field is unhindered, while in the plane perpendicular to the $\vec{B}$ field the projected path of the particle describes a circle of gyroradius

$$r_g = \frac{\gamma_\perp mc^2}{ZeB} = \frac{(E_\perp/eV)}{300Z(B/\text{gauss})}\,\text{cm}, \tag{10.1}$$

where $\gamma_\perp$ and $E_\perp$ refer to the Lorentz factor and the energy of the particle in the plane perpendicular to $\vec{B}$. As intuitively expected, the larger the magnetic field and the charge, the more tightly the charge is bound to the field and the smaller is the gyroradius, whereas the larger the particle energy, the less the field is able to bend its path and the larger is the gyroradius. For protons of energy $1\,\text{EeV} = 10^{18}\,\text{eV}$, the gyroradius in our galaxy's magnetic field of $B \sim 3 \times 10^{-6}$ gauss is $r_g \sim 1\,\text{kpc}$, while for the same proton in intergalactic space, where the fields are closer to $B \sim 10^{-9}$ gauss, the gyroradius is $r_g \sim 1\,\text{Mpc}$. Cosmic-ray protons are observed at energies ranging from $\gtrsim 10^9\,\text{eV} \equiv \text{GeV}$ up to $\sim 10^{20}\,\text{eV} \equiv 100\,\text{EeV}$. Such a cosmic ray, even if it is a proton of sub-microscopic mass $m_p = 1.67 \times 10^{-24}$ g, has a macroscopic kinetic energy of $\sim 1.5 \times 10^8$ erg $\sim$ 15 joules, comparable to the energy of a tennis ball traveling at 80 miles per hour. Cosmic rays in the range $E \gtrsim 0.1\,\text{EeV}$ are referred to as ultra-high energy cosmic rays.

If the size of the gyroradius is smaller than the size of an astrophysical system, say our galaxy, one can see how a cosmic ray would be trapped inside the system, as it is forced to turn around in circles smaller than the shortest escape route. However, from eq. (10.1) we see that for EeV energy cosmic rays, the gyroradii in the average Milky Way field of $B \sim 3\,\mu\text{G}$ is larger than the height of the galactic disk ($\sim 200\,\text{pc}$), and there is no way that cosmic rays of such energy or larger could be contained within our galaxy. They must have come from outside, from extragalactic distances.

On the other hand, cosmic rays of energy $\lesssim$ EeV could in principle originate from within our galaxy. The fact that we see them arrive isotropically, despite

---

[1] The force acting on a particle of charge e is given by the vector product $\vec{f} = e[(\vec{v}/c) \times \vec{B}]$.

the fact that the galaxy is disk shaped, can be understood if the cosmic rays have made so many circles that they have completely lost memory of where in the galaxy they originally came from.

In fact, cosmic rays of energy $\lesssim$ PeV $\equiv 10^{15}$ eV almost certainly come from within our galaxy, and we even have a good idea of where and how they achieved such energies. The where follows from a total source energy argument. We know from measurements what the average flux of cosmic rays is that hits the Earth in this energy range, which tells us what is the energy density (energy per unit volume) of such cosmic rays in the galaxy. We also have a good idea of how long it would take before such cosmic rays finally escaped from our galaxy, after many circles.

To see that cosmic rays must eventually escape any given system, even though their gyroradius is much smaller than the system size, in this case the galaxy, think of a whirling dervish, or a blindfolded dancer who dances in circles while moving in a given average direction. Note that the direction of the magnetic fields is random, with many changes of direction, like a zig-zag pattern. This is equivalent to a changing of the general direction of motion of the dancer every few circles. Thus, the dancer will eventually leave any given sized dance floor, it just takes some time before the random zig-zag plus circling finally gets the dancer to intersect one of the outer boundaries.

Now, knowing the cosmic-ray energy density and their average residence time in the galaxy, we know what the energy production rate in cosmic rays must be. There are very few sources in the galaxy with enough energy to supply this. The likeliest candidates are in fact supernovae, which occur at an average rate of one per 30 to 100 years in our galaxy, each injecting on average $\sim 10^{43}$ erg s^{-1} of kinetic energy into the galaxy. For a conservative $\sim 1\%$ efficiency of conversion of this energy into cosmic rays, this is sufficient to supply the required $\sim 5 \times 10^{40}$ erg s^{-1} in cosmic rays observed.

So how do supernovae accelerate protons to energies $E_p \lesssim 1$ PeV? The mechanism which is thought to be responsible is called Fermi acceleration, which occurs in the shock wave driven by the supernova into the interstellar medium surrounding it. The supernova explosion is a sudden release of a huge amount of energy, which results in the ejection of the outer envelope of the star. The velocity of the ejecta is large, even if sub-relativistic, $v/c \sim 0.1$, but it is much larger than the speed of sound in the interstellar medium into which it is advancing, resulting in a shock wave moving ahead of the ejecta.

Now, there are random magnetic fields both in the shocked gas downstream of the shock and in the unshocked interstellar medium (ISM) upstream of the shock. Considering, for example, the protons in the shocked gas, which are bouncing around in the random magnetic fields, there are some which by the laws of statistics are in the high velocity tail of the thermal velocity distribution, moving randomly much faster than the average bulk velocity of the shocked gas. These faster protons will reach the shock boundary faster, and can move from one side of the shock to the other. Once on the other side, they will quickly be deflected and bounced around by the magnetic fields in the ISM, and while still endowed with very fast random velocities, they will acquire the average bulk motion of the unshocked ISM, which is very different from that of the shocked gas region from which they came. This net bulk velocity difference represents at each crossing a net increase in the energy of the protons (i.e. cosmic rays). Then, after a few bounces, some of them will again randomly reach the shock boundary and make their way back into the shocked gas region, where they are quickly randomized again to acquire the bulk velocity of the shocked gas, again resulting in a net gain of energy because of the net difference in the two bulk velocities. In this manner, there is a smaller and smaller number of cosmic rays making a larger and larger number of transitions between shocked and unshocked gas, acquiring extra energy at each transition. The process resembles that of a particle bouncing between converging mirrors with a net relative approach velocity. At each bounce, the kick from the approaching mirror gives the particle extra energy.

How much energy can a cosmic ray acquire in this manner? One can intuitively guess that the energy acquired by the cosmic ray must be larger in some proportion to the size of the acceleration region (the shock radius R_s, say), the bouncing ability (i.e., the field strength B, which determines how sharp the deflection of the path will be), and the shock velocity $\beta_s = v_s/c$. It can be shown that the energy acquired by the particle is linearly proportional to these quantities, as well as the electric charge,

$$E_p \lesssim \beta_s Z e B R. \tag{10.2}$$

There is a natural maximum energy possible for a cosmic ray accelerated in this manner, which is reached when the cosmic ray gyroradius [eq. (10.1)] equals the size of the shock radius or the ejecta radius R_s, since at that energy the cosmic ray can no longer be contained (i.e., bounce) inside the acceleration region. Putting in typical numbers, for $Z = 1$ (protons) in a supernova remnant this energy turns out to be $E_{p,SN} \sim 10^{15}$ eV $\equiv$ PeV, furnished with which the cosmic ray then escapes into the relative freedom of the galaxy. For heavier nuclei (e.g., iron; $Z = 26$), the maximum energy is correspondingly larger, and in fact it is thought that up to $E \lesssim 10^{17}$ eV the composition of the observed cosmic rays becomes richer in heavy nuclei, and normal supernovae may be responsible for galactic cosmic rays up to this energy.

10.2   Ultra-high energy cosmic rays

Ultra-high energy cosmic rays, or UHECR for short, is the generic name for cosmic rays in the range $E \gtrsim 10^{17}$ eV, whose origin is less clear. It is possible that supernovae resulting from the collapse of stars which had a substantial stellar wind before the explosion result in larger shock radii. Also, there is a sub-class of core collapse supernovae called hypernovae (see Chapter 7), which appear endowed with larger ejecta velocities and larger energies. In both cases, it may be possible to get cosmic nuclei up to $E \sim 10^{18}$ eV still within our galaxy. However, as discussed above, cosmic rays of energies larger than this must arise outside our galaxy, and two major questions offer themselves: are there any higher energy cosmic rays beyond the currently observed maximum values $\sim$ few $\times 10^{20}$ eV? and what is capable of accelerating them up to these huge energies?

The first question has a natural theoretical answer, which after 40 years appears to receive support from observations. This is the so-called GZK effect, named after Greisen, Zatsepin and Kuzmin. This was formulated soon after the discovery of the cosmic microwave background (CMB) in 1965, and it is based on a well-tested phenomenon of nuclear physics, which is the photo-meson effect. It is observed in the lab that if photons ϵ are incident on a proton, and if the energy of the photon as seen by the proton exceeds somewhat the rest-mass energy of a pion, the collision results in the production of new particles, namely either a neutral pion plus a weakened proton, or a positively charged pion plus a weaker neutron,

$$p + \epsilon \rightarrow p' + \pi^0,$$
$$p + \epsilon \rightarrow n + \pi^+. \tag{10.3}$$

The mass of the pion is $\sim 140 \,\text{MeV}/c^2$, so the microwave background photons, whose mean energy is $\epsilon \sim 10^{-3}$ eV, would in principle seem not to pose a danger of this happening. However, for protons whose energy is $E_p \sim 10^{20}$ eV with a Lorentz factor $\gamma_p \sim E_p/m_p c^2 \sim 10^{20} \,\text{eV}/10^9 \,\text{eV} \sim 10^{11}$, the puny little CMB photon of mean energy 10^{-3} eV appears to be much more energetic, being boosted in the proton rest frame by the Lorentz factor to an energy $E_{CMB,rest} \sim 10^{-3} \times 10^{11} \sim 10^8$ eV ~ 100 MeV, close to the pion rest-mass energy. Since the CMB photons have a black-body distribution extending both below and above the mean energy, there is a sufficient number of photons whose energy in the proton rest frame is enough to produce pions. The energy needed for creating the pion comes at the expense of the proton, which consequently loses energy. Thus, if a proton had initially an energy in excess of this GZK energy,

$$E_{GZK} \sim 10^{20} \,\text{eV}, \tag{10.4}$$

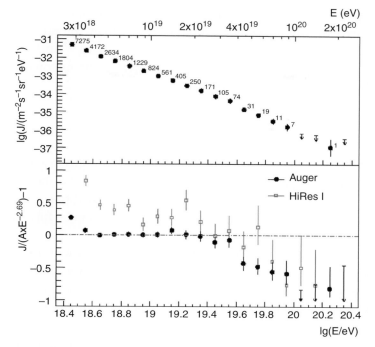

Figure 10.1   Cosmic-ray spectral flux detected with Auger [58].
Source: Copyright 2008 by The American Physical Society.

it would quickly lose energy by interactions with CMB photons, until its energy drops below E_{GZK}, which is the threshold energy for pion production. Since 2007, observations with the Pierre Auger cosmic ray observatory have begun to show that there appears indeed to be a steepening of the cosmic-ray spectrum beyond $\sim 6 \times 10^{19}$ eV, which appears compatible with the expectations from the GZK effect (see Fig. 10.1).

This still does not prove that protons cannot be accelerated to energies beyond E_{GZK}: it merely proves that a proton with energy above the GZK values will lose its energy after it encounters enough CMB photons to degrade its energy, at the rate of ~ 140 MeV per encounter. The mean free path[2] of a proton between collisions with photons is $\lambda \sim 1/(\sigma_{p\gamma}.n_{\gamma,th})$, that is, the smaller the density of targets and the smaller the probability of interaction per encounter, the longer the proton can go without suffering a photo-meson interaction. The spatial density of "average" CMB photons is measured to be $n_\gamma \sim 380$ cm^{-3}, and

---

[2] The concept of mean free path was discussed in Section 9.1 for photon collisions with other photons.

the density of photons in the high energy tail near threshold $n_{\gamma,th}$ is roughly 15–20 times less, while the cross-section of the photo-meson process (roughly the probability of interaction per encounter) is $\sigma_{p\gamma} \sim 5 \times 10^{-28}$ cm^2. Using these numbers, the mean free path λ for protons subject to photo-meson interactions against CMB photons defines a "GZK radius" around us,

$$R_{GZK} \sim \frac{1}{\sigma_{p\gamma} n_{\gamma,th}} \sim 50\,\text{Mpc}. \tag{10.5}$$

If there are sources which are capable of accelerating protons to energies much higher than E_{GZK}, eq. (10.5) says that if these sources are at distances from us larger than R_{GZK} their energies would have been degraded to $E \lesssim E_{GZK}$. Thus, any sources producing photons at energies $E \sim E_{GZK}$ cannot be further from us than $R_{GZK} \sim 50$ Mpc [59].

From eq. (10.1) it is seen that the gyroradius increases with the energy of the particle. However, the scrambling or isotropization of the arrival directions of an ultra-high energy cosmic ray in a random magnetic field decreases with energy. This is because the magnetic fields in intergalactic space are relatively weak and randomly oriented, with many twists between a source at distance R_{GZK} and us, and the gyroradius of a cosmic ray with GZK energy is much larger than the typical length scale λ over which the magnetic field changes direction. Thus, the cosmic ray just glides over the small ripples in the field, and the larger its gyroradius (the larger its energy) the less it feels the ripples. This is similar to the reason why a large ship is less sensitive to the effects of the ocean waves and just forges ahead majestically, while a smaller ship is buffeted around by the waves. The end result is that the angle of deflection of the cosmic ray varies proportionally to the magnetic field strength, as expected, but inversely proportionally to the cosmic-ray energy.[3] Putting in numbers for the intergalactic magnetic field strength and an estimated magnetic field coherence

---

[3] This can be seen by considering that for a magnetic field which is more or less straight over a distance λ before bending, the cosmic ray while gliding down this straight stretch of field line gyrates around it with a gyroradius r_g, which for intergalactic fields and UHECR energies is much larger than the typical field coherence length λ. Thus, after traveling a path λ, the cosmic ray has been deflected by a small angle $\theta_\lambda = \sin^{-1}(\lambda/r_g) \sim (\lambda/r_g) \ll 1$. Thus the angle of deflection is *smaller* for higher energies, since $\theta_\lambda \propto r_g^{-1} \propto E^{-1}$. For a randomly oriented magnetic field which changes its direction on average after a distance λ traveled, if the proton arrives after n such changes of direction it can be seen that the net final angle cannot be simply $n\theta_\lambda$, since the random changes of direction also imply that the net direction of deflection changes. It can be shown that the net angular displacement from its original direction will be $\theta \sim n^{1/2}\theta_\lambda$. Here $n \sim r/\lambda$ increases with the total distance r traveled, but the net angular deflection will still be inversely proportional to the particle energy.

length $\lambda \sim 1\,\text{Mpc}$, it turns out that after traveling 100 Mpc, UHECR of $E \sim 1\,\text{EeV}$ would be very strongly scrambled ($\theta \sim 1$ radian), but UHECR of $E \sim 10^{20}\,\text{EeV}$ would only be scrambled by a net angle of $\theta \sim 10^{-2}$ radian ~ 0.5 degrees. That is, protons of GZK energy would arrive to us from their source (which should be within $R_{GZK} \lesssim 50\,\text{Mpc}$) pointing back to their source with fairly good accuracy, at least by cosmic ray standards. Of course, this angular resolution is far below the typical optical astronomy angular resolution of less than an arc-second (1/3600 of a degree), but even so, there are not many sources, if any, within distances of $\lesssim 100\,\text{Mpc}$ within a $1°$ angular cone which could be suspected of being capable of producing such cosmic rays.

Thus, we know that a GZK cosmic ray has traveled a distance of at most $D \sim$ few $R_{GZK} \lesssim 100\,\text{Mpc}$, and it must have come from some object which *was* located within about a degree of the direction of arrival of the UHECR. We say *was* because even traveling on a straight line the ultra-relativistic particle would take a time $t = D/c \simeq 3 \times 10^8$ yr to arrive here, with a random time delay due to zig-zags providing a very small additional value of $\Delta t \sim 10^3$–10^4 yr for a GZK energy proton. This delay is proportional to $(\lambda B^2 D^2 / E^2)$.

But what could have accelerated it to such high energies? Supernovae, we know, can at most get iron up to 10^{17} eV, and anomalous supernovae or hypernovae at best would get us to 10^{18} eV or so. The basic acceleration limit is given by eq. (10.2), setting R equal to the size of the acceleration shock region endowed with an average magnetic field B. One can search the astrophysical zoo for objects with large values of the product of R and B, and one comes up with two likeliest candidates: active galactic nuclei with powerful jets (AGN) and gamma-ray bursts (GRB).

The two most exciting events of 2007 in this field were the publication of the first results from the Pierre Auger Cosmic Ray Observatory on the spectrum of UHECR [59] (showing a decline at energies $E \gtrsim E_{GZK}$) and on the spatial (angular) distribution of UHECR [60]. The latter result used UHECR collected in the first year of operation and correlated their directions of arrival against AGNs from an AGN catalog. They used a particular set of AGNs (from the Veron–Cetty catalog), and varied the lower energy limit of the UHECR E, the allowed angular deflection circle θ_{AGN} around an AGN, and the maximum distance D_{AGN} considered for the correlation, and found a maximum correlation when taking the 27 UHECR collected at $E \geq 6 \times 10^{19}$ eV with AGN within $\theta_{AGN} = 3°$ within 75 Mpc from Earth. This maximum correlation was found to have a probability of 99.3%, or in statistical jargon, to be within 2.7σ standard deviations from a random result. This is not a very strong probability, by physics standards: one normally requires a probability of 99.999996%, or 5σ to accept a result as being beyond reasonable doubt, but still, a 2.7σ result is suggestive. It was announced with due cautionary reserve by the Auger group, and reviewed with much more enthusiasm by the semi-popular press. Unfortunately, a subsequent reanalysis by the Auger group a year later reduced the probability to 95%, or $\lesssim 2\sigma$, and the question must be considered open.

However, there is one conclusion from the Auger result which they stressed appears quite safe: the UHECR definitely correlate with the concentrations of matter, that is, the large-scale structure of galaxies. GRBs occur in galaxies at a rate of roughly one every 10^4–10^6 yr per galaxy, depending on the galaxy type, and there are roughly 10^5 galaxies within a sphere of 100 Mpc radius. We cannot be sure whether we have detected a GRB in the last year within 100 Mpc, since most GRBs do not have a redshift distance determination. The average rate of occurrence would indicate about one per year. However, if we consider the UHECR arrival time delay $\Delta t \sim 10^4$ yr for UHECR with $\sim 6 \times 10^{19}$ eV from that distance, the UHECR arriving now would have originated in 10^3–10^4 GRBs occurring within a comparable period of 10^3–10^4 yr, ensuring that their arrival directions would be isotropic and biased towards the galaxy mass concentrations. This is based on assuming that the GRBs emit an amount of energy in 10^{19}–10^{21} eV cosmic rays comparable to or maybe a factor ~ 10 greater than the energy they emit in observed gamma-rays [61], a physically plausible assumption.

10.3   Cosmic-ray observational techniques

The cosmic rays incident on the top of the Earth's atmosphere suffer a first interaction with an atmospheric nitrogen or oxygen atom, which results in the production of energetic charged and neutral secondaries, which in turn interact again and make tertiary particles, etc., resulting in a shower of particles which propagate downward into the atmosphere (Fig. 10.2). The cascade consists of two types of processes, the hadronic and the electromagnetic cascades. An incident nucleon will lead to both $\pi^\pm$ and π^0 in the approximate ratio of 2 to 1, and up to the highest CR energies, the π^0 decay into two photons in a time shorter than the time needed to interact again. These photons then produce e^+e^- pairs, which produce more photons, etc. Similarly, an incident high energy photon will also produce pairs, which produce photons, which produce more pairs, etc., as we discussed for TeV photon observations in Chapter 8. In both cases, an *electromagnetic* cascade ensues. In the incident nucleon cases, however, the weakened nucleon as well as the produced charged pions, kaons, etc. continue to interact via strong interactions, producing an increasing number of strongly interacting secondary particles in a *hadronic* cascade. In both types of cascades, the initial particle loses a significant fraction of its energy and

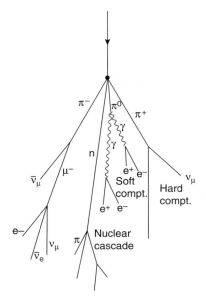

Figure 10.2 Schematic production and development of a cosmic-ray shower, showing the soft electromagnetic, the hard penetrating muon, and the hadronic components.

Source: The Auger group.

produces one or more new particles, which split among themselves the energy lost by the parent particle. Thus an incident (nucleon) cosmic-ray cascade has an electromagnetic, a muonic and a hadronic component.

The hadronic component exhausts itself in the atmosphere, converting all of its energy into muons and an electromagnetic cascade. For UHECR typically the muons reach the ground, where they are measured with surface detectors (SD). Being electrically charged this can be done using, for example, scintillation counters as in the AGASA experiment, or via the Cherenkov light they emit in water tanks, as implemented in the Auger experiment (shown schematically in Fig. 10.3). The number of electrons and positrons in the electromagnetic cascade multiplies and reaches a maximum somewhere at mid-atmospheric levels. These $e^{\pm}$ excite atmospheric nitrogen atoms, which emit fluorescent light that can be detected by means of optical telescopes (fluorescence detectors, or FD). The latter are arranged as batteries of telescopes covering a wide field of view in a so-called Fly's Eye arrangement. The latter is named after an earlier experiment in Utah (the current improved experiment is called Hi-Res), and it refers to the eyes of a fly being made up of many detectors, each slightly off-set from its neighbors so as to cover a wide field of view. The Auger experiment

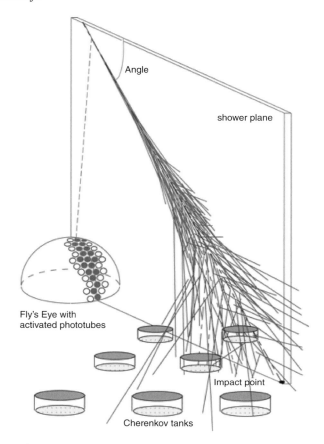

Figure 10.3 Auger hybrid detector scheme, showing atmospheric shower development and observation with a Fly's Eye type fluorescence detector and water Cherenkov surface detectors.
Source: The Auger group.

uses this Fly's Eye FD technique (Fig. 10.3) combined with the muon surface detectors, its strength being that it is a hybrid detector combining the AGASA and Hi-Res techniques, allowing for better cross-calibration.

The Pierre Auger Cosmic Ray Observatory is located in the foothills of the Andes in Argentina, and consists of 1600 water Cherenkov tanks to measure muons reaching the ground, distributed over a total area of 3000 km^2. One of these is shown in Fig. 10.4. In addition, Auger also has 24 fluorescence detector telescopes looking at the sky, so as to cover the air space above the area covered by the surface detectors.

The current energy sensitivity of Auger extends from $10^{17.5}$ eV to $\sim 10^{21}$ eV, with a planned extension called HEAT-AMIGA which is designed to reach down

Figure 10.4   One of the 1600 water Cherenkov tanks of the SD system of the
Pierre Auger Cosmic Ray Observatory.
Source: The Auger group.

to $\sim 10^{16}$ eV. One of the other capabilities of cosmic-ray detector arrays is that
they can provide some information on the chemical composition of the arriving UHECR. With the fluorescence technique this exploits the fact that heavier
nuclei initiate cascades higher up in the atmosphere, while with the surface
detectors one exploits the fact that heavy nuclei result in more muon-rich showers relative to electrons. Also, in general, proton showers show more variability
than heavy nucleon showers. The current results indicate that at GZK energies
there appears to be a contribution from heavier nuclei, but the exact fraction
and atomic weight is still uncertain.

11

Neutrinos

11.1   The elusive neutrinos

Neutrinos are the most elusive of all the known particles. Their existence was originally postulated by Wolfgang Pauli in order to explain a small amount of energy and momentum which appeared to be missing following certain nuclear interactions, but could be accounted for if there was a very light, hard-to-detect particle that carried away unseen the missing energy–momentum. In the mid-1930s Enrico Fermi then developed a theory of weak interactions which became an integral part of nuclear physics, but it was not until the mid-1950s that Reines and Cowan actually detected neutrinos experimentally, proving their existence [10]. They did this with anti-neutrinos $\bar{\nu}_e$ produced in a nuclear reactor, which were allowed to interact with protons from water in a large underground container to produce a neutron and a positron,

$$\bar{\nu}_e + p^+ \rightarrow n + e^+, \tag{11.1}$$

which is a typical weak interaction process. The positrons were then detected through the two gamma-rays produced when they annihilated against electrons in the water, while a short time later the neutrons were captured by another nucleus emitting another gamma-ray, in an unmistakable pattern.

The cross-section σ (i.e., the probability of interaction per second of a given target proton divided by the incident flux of anti-neutrinos per unit area per second) is extremely small, of order 6×10^{-44} cm^2 for an MeV energy experiment such as eq. (11.1), about 20 orders of magnitude smaller than the cross-section for the more mundane electromagnetic process of light being scattered by an electron. Thus, neutrinos are extremely hard to detect, since they can pass

Figure 11.1   The Davis solar neutrino experiment in the Homestake mine.
Source: Courtesy of Brookhaven National Laboratory.

right through even huge column densities of material with only an infinitesimal chance of interacting. The nuclear reactions in the Sun lead to 6.5×10^{10} (65 billion) neutrinos per square centimeter per second incident on Earth, which pass through our bodies essentially unnoticed. It took Davis years of letting these solar neutrinos pass through 100 000 gallons (400 kiloliters) of the common cleaning fluid perchlorethylene, until he accumulated enough neutrino interactions with the chlorine to actually measure the flux of solar neutrinos at Earth, essentially in agreement with the solar models calculated by Bahcall. See Fig. 11.1.

Actually, the flux of electron neutrinos ν_e measured by Davis was about 40% lower than what was predicted to be produced by the nuclear reactions in the Sun. However, a solution emerged to this discrepancy, which was developed through the 1980s and 1990s, namely, that a fraction of the electron neutrinos on their way out from the Sun's core oscillate over into a different flavor of neutrinos, muon neutrinos ν_μ (see Chapter 2). The presence and the amounts of muon neutrinos attributed to such oscillations were measured by the Super-Kamiokande experiment in Japan, the SNO experiment in Canada, and several other experiments.

This phenomenon of neutrino oscillations provided strong experimental proof of the existence of physics beyond the Standard Model of particle physics. It is intimately linked to the fact that neutrinos have a very small but non-zero mass. Other experiments using atmospheric neutrinos (neutrinos produced when cosmic rays interact with the Earth's atmosphere) also showed the phenomenon of oscillation, as did experiments done with underground detectors measuring neutrinos from the beam of a distant particle accelerator. Ray Davis and Masatoshi Koshiba received the physics Nobel prize in 2002 for their pioneering experiments in this area [62].

11.2   Stellar and supernova neutrinos

Stars, by definition, are large mass agglomerations undergoing nuclear fusion reactions at their center, as opposed to planets and smaller objects which do not. Nuclear fusion processes are accompanied by neutrino emission, and typically such stellar neutrinos have roughly thermal energy distributions with a mean energy of $\sim 1-30$ MeV. These are called thermal neutrinos, since their energies are comparable to the kinetic temperature of the stellar interior, which is set by a balance between gravity and the opposing thermal pressure. When massive stars undergo core collapse leading to supernovae (which are called Type II, Ib or Ic), or when white dwarfs accrete enough matter to initiate carbon ignition leading to a supernova (in this case called Type Ia), the nuclear processes involved also produce thermal neutrinos of energy $\sim 10-30$ MeV.

In this thermal energy range of order MeV, the neutrino cross-sections for interacting with matter are on the order of 10^{-44} cm^2 and thus extremely small, requiring huge detector volumes, such as those employed in the chlorine experiment in Homestake, South Dakota in the USA, the Kamiokande water experiment in Japan (Fig. 11.2), the IMB experiment and others (see Section 11.6). These experiments have so far detected thermal neutrinos from two relatively nearby stellar sources: the Sun, at a distance of 1.5×10^8 km $= 1.5 \times 10^{13}$ cm, and the supernova SN 1987a, which occurred in the Large Magellanic Cloud, about 55 Mpc $\simeq 1.6 \times 10^{26}$ cm away from us. Needless to say, we receive many more neutrinos from the Sun than we did from SN 1987a. Still, a supernova explosion is a one-shot affair which delivers in one jolt an energy comparable to that which the star previously dribbled out over billions of years, so even at that distance we did observe 24 neutrinos from SN 1987a, with three different detectors [63]. The time spread of the neutrinos was about 13 seconds, which is the expected duration for the neutrinos from a core collapse to form

Figure 11.2   The new Super-Kamiokande, successor of the Kamiokande detector in a mine in Japan, 1000 m (3281 ft) underground, showing the spherical phototube arrays being serviced.
Source: Kamioka Observatory, ICRR (Institute for Cosmic Ray Research), The University of Tokyo.

and escape the core. One of the detector groups claimed a second batch of three neutrinos two hours after the first batch, which prompted speculation that the first main neutrino burst may have been the collapse to a neutron star, followed by a second collapse to a black hole two hours later, but the reality of the second burst has not been confirmed.

The role of neutrino oscillations in stellar processes is largely confined to electrons and muon neutrinos. It is mostly electron neutrinos or anti-neutrinos which are produced, since the temperatures are mostly well below the 105 MeV rest-mass of the muon, and oscillations are responsible for the appearance of the muon neutrino flavor, with tau neutrinos playing a minimal role in the stellar case. In the dense gaseous environment of the stellar interior, whether in steady burning or in supernovae, the presence of matter can lead to an increase of the oscillation probability when the neutrinos go through regions of just the right density, which leads to enhanced flavor conversion over what would have occurred in vacuum. This phenomenon is referred to as matter oscillations, or the MSW effect, named after Mikheyev, Smirnov and Wolfenstein.

11.3   Atmospheric neutrinos

Cosmic-ray nucleons which permeate the Universe regularly slam into the Earth's atmosphere and produce neutrinos, in processes such as

$$p^+ + p^+ \rightarrow \pi^+ + p^+ + n,$$
$$p^+ + n \;\; \rightarrow \pi^+ + n + n, \tag{11.2}$$

followed by

$$\pi^+ \rightarrow \mu^+ + \nu_\mu \rightarrow e^+ + \nu_e + \bar{\nu}_\mu, \tag{11.3}$$

and other similar processes, resulting in both muon and electron neutrinos. These neutrinos are more energetic than the stellar ones, depending on the triggering cosmic-ray energy, most observations being in the hundreds of MeV to tens of GeV. Since the cosmic-ray spectrum decays as a steep power law of energy, $N(E) \propto E^{-2.7}$ below PeV energies, the corresponding atmospheric neutrinos are also much more abundant at lower energies. This results in downward-moving neutrinos from cosmic rays arriving in the atmosphere above the detector, and upward-moving neutrinos from cosmic rays arriving from the opposite terrestrial hemisphere, the resulting neutrinos coming right up through the Earth, which is essentially transparent to them at energies $\lesssim$ PeV.

The fact that atmospheric neutrinos arriving from different angles have traversed different column densities of matter is particularly useful for measuring the neutrino oscillation phenomenon. The neutrino flux measured at the detector in different flavors shows a marked angular dependence, departing from the simple 2:1 muon to electron flavor ratios expected from the production mechanism [eq. (11.3)]. This is because at different angles of arrival the neutrinos (and muons) have traversed different column densities of matter, resulting in different amounts of redistribution among the flavors, including in this case also oscillation into the tau neutrino flavor.

This type of experiment, combined with solar neutrino oscillation experiments and other accelerator or reactor-based experiments, provides information on the difference of squares of the neutrino masses. These, combined with other laboratory and cosmological experiments, have provided upper limits on the masses of individual neutrino flavors. The current limits from laboratory data give the individual limits $m_{\nu_e} < 2.2\,\text{eV}$, $m_{\nu_\mu} < 170\,\text{keV}$, $m_{\nu_\tau} < 15.5\,\text{MeV}$, while cosmological data provide a stronger upper limit that the sum of the masses over all three flavors is [11] $\sum_i^3 m_{\nu_i} < 0.7\,\text{eV}$.

Since the density and chemical composition of the Earth vary with depth, this is also a valuable tool for probing the Earth's mantle and core, cross-checking against geophysical data from other techniques such as seismology, etc.

11.4   VHE astrophysical neutrinos

Another process which produces neutrinos is called the photo-meson process, which occurs when high energy nucleons interact with photons, provided that the total energy of the two particles is above a minimum threshold value. For a proton (e.g., a cosmic ray) interacting with "soft" photons γ_s this process leads to

$$p^+ + \gamma_s \to \begin{cases} \pi^+ + n & \to \mu^+ + \nu_\mu \to e^+ + \nu_e + \bar{\nu}_\mu \\ \pi^0 + p^+ & \to 2\gamma \end{cases} \tag{11.4}$$

where in the second line the π^0 decay results in two "hard" gamma-rays of $\gtrsim 70\,\text{MeV}$ each. The threshold proton energy for this process against target photons γ_s of energy ϵ_s is approximately (ignoring angle factors) given by requiring that the geometric mean of the initial proton and photon energies equals that of the final proton and a pion,[1]

$$\left(E_p^{thr}\epsilon_s\right)^{1/2} \sim \left(m_p m_\pi\right)^{1/2} \sim 0.35\,\text{GeV}. \tag{11.5}$$

In the π^+ decay, the neutrino energies are lower than the parent proton energies by a factor $\sim 1/20$. This is because the average energy of the pion is $\sim 1/5$ of the energy of the parent proton, and the (charged) pion decay chain $\pi^+ \to \mu^+ + \nu_\mu \to e^+ \nu_e \bar{\nu}_\mu$ results in four leptons, each of which roughly carries 1/4 of the pion energy. On the other hand, the branching ratio for the process $p\gamma \to p\pi^0$ is about twice as large as that for $p\gamma \to p\pi^+$, while the energy of the π^0 is slightly smaller, so that the energy lost to neutrinos is $(1/3) \times (3/4) = 1/4$ (because 1/4 goes into a positron), while that lost to photons is $(2/3) + (1/3) \times (1/4) = 3/4$ (because the positron eventually leads to photons). Thus the ratio of neutrino to photon luminosity is $L_\nu : L_\gamma = 1 : 3$. There is in addition some proton energy

---

[1] More accurately, the threshold energy is $E_p^{thr}\epsilon_s \geq \frac{m_\pi c^4(2m_p+m_\pi)}{2(1-\cos\theta)} \sim \frac{0.13\,\text{GeV}^2}{(1-\cos\theta)}$, where θ is the relative angle between proton and photon, and in the second term the energies are expressed in GeV.

which is lost directly to pair production via $p\gamma \to e^+e^-p$, so in the end one has roughly $L_\nu \sim (3/13)L_\gamma$, if π^0 decay is the only source of observed gamma-rays. However, the condition (11.5) implies that the target photons γ_s for the photo-meson process must be much softer than the gamma-rays resulting from the π^0 decay, that is, they must be some other ambient photons.

The photo-meson process is expected in all sources which accelerate cosmic rays, most prominently, in supernova remnants, active galactic nuclei and gamma-ray bursts, but it may occur also in micro-quasars and possibly pulsars or magnetars. The importance of the effect is of course dependent also on an adequate column density of photons satisfying the threshold condition [64].

Supernova remnants are the prime candidate sources for accelerating the cosmic rays observed up to $E_p \sim 10^{15}$ eV, and one expects pp and pn collisions of cosmic rays with thermal nucleons in the remnant shell, leading to TeV neutrinos from π^+ decay and TeV gamma-rays from π^0 decay. Even though there is a good number of SNRs in our galaxy which are observed as TeV gamma-ray sources, if these are due to pp and pn collisions the neutrinos are not expected to be detectable, as these are much harder to detect than TeV gamma-rays. In addition, there is controversy as to what fraction of the TeV gamma-rays are due to π^0 decay, since inverse Compton scattering of microwave and infrared photons can also reproduce the observed TeV photon spectra (see Chapter 8). The latter interpretation has its problems, but so does the π^0 decay, and this question remains unsolved.

Active galactic nuclei (AGNs) are also observed to accelerate electrons in shocks either inside their jets (Fig. 11.3) or at the jet termination surface, and as discussed in Chapter 10, certain classes of AGN may be capable of accelerating cosmic rays up to 10^{20} eV (GZK) energies. A larger subset of AGNs should be able to accelerate cosmic rays to energies in excess of PeV, which could lead to TeV neutrinos and gamma-rays via photo-meson processes, given the appropriate target photon column densities. As discussed in Chapter 8, the two-hump photon spectra of blazars are thought to be due to electron synchrotron for the lower hump peaking at UV to X-ray energies, and due to either electron inverse Compton or π^0 decay for the higher hump peaking at GeV to TeV energies. However, there is so far no conclusive evidence ruling out one or the other mechanism, and again the possibility exists that both may be present, at a relative fractional level which is uncertain. In AGNs, however, although they are at extragalactic distances the fluxes are sufficiently high that, if the hadronic (i.e., photo-meson) interpretation is correct, there is serious hope of measuring the corresponding TeV neutrino fluxes. This would be a "smoking gun" proof of the hadronic interpretation, and also of the acceleration of cosmic rays to at least PeV energies in these sources.

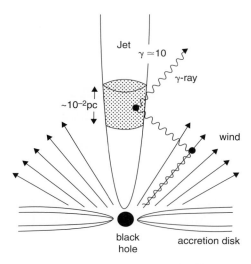

Figure 11.3 Schematic blazar internal shock scenario leading to $p\gamma \to \nu$. *Source*: F. Halzen.

Gamma-ray bursts are also thought to accelerate electrons to relativistic energies in shocks within and at the end of highly relativistic jets, resulting in intense MeV gamma-ray, X-ray and optical photon fluxes (Chapter 7). These shocks are also prime candidates for the acceleration of cosmic-ray protons, all the way up to GZK energies (Chapter 10). GRBs have been observed at tens of GeV photon energies with AGILE and Fermi, but since mostly they are at redshifts $z \gtrsim 1$, TeV gamma-ray searches have so far been fruitless, unsurprisingly since as discussed in Chapter 8, TeV photons would annihilate against infrared diffuse background photons over distances in excess of about 100 Mpc. For the GRB jets of bulk Lorentz factor $\Gamma \sim 10^{2.5}$, the threshold condition (11.5) is expressed as $\epsilon_s E_p^{thr} \gtrsim 0.3\,\mathrm{GeV}^2\Gamma^2$, which means that their gamma-ray photons are distributed as a broken power law with an MeV spectral break resulting in a neutrino spectrum which is also a broken power law extending below and above a neutrino spectral break at around 100 TeV (curve marked "burst" in Fig. 11.4). In addition, the UV photons expected from the reverse external shock would interact with GZK energy protons, resulting in a separate neutrino spectral component extending up to EeV neutrino energies (the component marked "afterglow" in Fig. 11.4). Whereas UHECR from GRBs are constrained to distances $d < D_{GZK} \sim 100$ Mpc, neutrinos have no such limitation – the Universe is transparent to them.

Other possible VHE neutrino sources are jet X-ray binaries, such as the microquasars discussed in Chapters 6, 8. These are suspected to harbor a central stellar

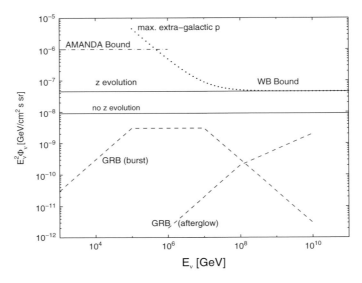

Figure 11.4   Diffuse neutrino spectrum from GRB models in units of energy per decade. Also shown are the IceCube sensitivity and the WB limit.
Source: E. Waxman.

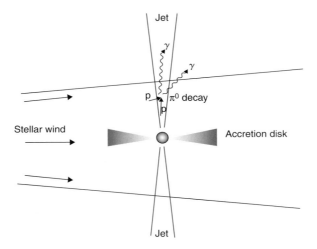

Figure 11.5   Schematic model of a possible hadronic model of micro-quasars. Both $pp \to \pi^0 \to 2\gamma$ and $p\gamma \to \pi^+ \to \nu$ are expected [65].

mass black hole, so they bear analogy with GRBs, the difference being that in micro-quasars there is a binary stellar companion which feeds the black hole matter in a more continuous, if irregular, manner. A better analogy is that of a scaled-down blazar, the difference in the origin of the accreted matter being irrelevant once an accretion disk forms around the black hole. In the case of micro-quasars, the stellar wind from the companion can also play a significant role, providing a source of both protons and photons for pp and $p\gamma$ processes in the jet, resulting in both TeV gamma-rays and TeV neutrinos (Fig. 11.5).

11.5   Cosmogenic neutrinos

Neutrinos at energies reaching up to $E_\nu \sim 5 \times 10^{18}$ eV can be expected from any sources which accelerate protons up to GZK energies of $E_p \sim 10^{20}$ eV. The best candidate sources for this are AGNs and GRBs, although hypernovae, magnetars and intergalactic shocks have been considered as well, with varying degrees of success. Of course, independently of what the sources are, the observed diffuse cosmic-ray spectrum must interact with any radiation fields in the cosmos, and it will undergo $p\gamma$ interactions contributing to both a diffuse gamma-ray background and a diffuse neutrino background.

PeV–EeV neutrinos are in fact expected from GRBs, as discussed above. For a source to photo-produce neutrinos up to 5 EeV it is necessary not only that it accelerate protons up to GZK energies, but it must also have a large enough photon target density of the appropriate energy for interacting [eq. (11.5)].

In any case, whatever the sources, there is a guaranteed source of EeV neutrinos, due to the observed diffuse cosmic-ray background. The UHECR spectrum in the range $10^{18}-10^{21}$ eV is almost certainly extragalactic, and if its origin is astrophysical, as appears to be the case, we must consider also their interactions before they leave the source, as well as after. The interactions of such cosmic rays with photons within the sources themselves typically lead to neutrinos which carry away a fraction $\eta_\nu \sim 0.15$ of the cosmic-ray energy, over all three types of neutrinos. Assuming a maximum efficiency $\eta_\nu = 1$ leads to an absolute upper bound on the diffuse very high energy neutrino flux, called the Waxman–Bahcall (WB) bound [66], also shown in Fig. 11.4.

There is an additional effect, which is due to the interactions of UHECR during their propagation through the Universe, in the course of which they interact with cosmic microwave background (CMB) photons, as well as with other background photons such as the diffuse infrared radiation background from star formation. The CMB and the $p\gamma$ photo-meson threshold energy is given approximately by eq. (11.5) for photons of energy ϵ. For the average CMB energy $\langle \epsilon_{CMB} \rangle = 6.34 \times 10^{-4}$ eV this threshold is $E_p^{thr} \simeq 2 \times 10^{20}$ eV. However, the UHECR can also interact with the high energy (Wien) tail of the CMB black-body distribution, so the real threshold is near 3×10^{19} eV. These interactions produce neutrinos, $p\gamma \rightarrow \pi+ \rightarrow \nu_\mu$, which are guaranteed to exist, since both the UHECR and the CMB are well measured, and the interaction is well understood.

The resulting neutrinos are typically at an energy $\sim 1/20$ of the original proton energy, so one expects this cosmogenic neutrino spectrum to be prominent in the EeV range.

However, the UHECR are measured only from sources within $D_{GZK} \lesssim 50-100$ Mpc, whereas the neutrinos arrive to us from much farther, all the way out to the Hubble horizon. The UHECR sources, whether they are AGN, GRB or whatever, must evolve with redshift, meaning that their luminosity and spatial density are different at different redshifts [67]. This evolution is known to occur for specific types of sources, although observationally they are not well determined. It is known to be different for the average star formation, for GRB, and for individual types of AGN. Thus, what is done is to take from observations the best guess for the redshift evolution law of different potential UHECR sources, and taking a standard cosmological model (say one with Hubble constant $H_o = 70$ km s^{-1} Mpc^{-1}, vacuum energy density $\Omega_\Lambda = 0.75$ and flat overall curvature $\Omega_{tot} = 1$) one proceeds to calculate a predicted cosmogenic neutrino spectral flux. These neutrino flux spectral distributions and intensities are different for different UHECR sources, and it is hoped that this will serve as a probe for identifying the real UHECR sources.

11.6   Neutrino detectors

High energy neutrinos are easier to detect than MeV neutrinos, since the neutrino–nucleon interaction cross-section increases with neutrino energy. On the other hand, the source fluxes decay steeply with energy, typically as $N(E_\nu) \propto E_\nu^{-2}$ or steeper, so very few high energy neutrinos are expected per unit detector area, and very large detectors are needed [68]. This is the basic reason for building km^3 scale detectors, such as IceCube under the Antarctic polar cap ice, or KM3NeT under the Mediterranean sea. The detection technique for $\nu_\mu, \bar{\nu}_\mu$ relies on the relativistic muons produced when the ν_μ interact with the target material or its surroundings, which then produce Cherenkov (optical) light. The light is distributed over a conical surface whose axis is along the muon path, and its properties determine the muon Lorentz factor (i.e., the parent neutrino's energy, see Fig. 11.6). Electron and tau neutrinos are also detected, but these produce cascades whose measurement technique is less straightforward.

The Cherenkov light is measured, if the medium is clear enough, with phototubes lowered by strings or attached to towers in the detector medium (ice or water). See Fig. 11.7. Both ice and water are fairly transparent, with typical photon scattering and absorption mean free paths of tens of meters. The scattering length is shorter/longer while the absorption length is longer/shorter for ice/water. The scattering mean free path determines the angular resolution

Figure 11.6 Schematic diagram of Cherenkov light from a muon caused by a muon neutrino in IceCube.
Source: IceCube collaboration/NSF.

(how well one can determine the direction of arrival of the neutrino), while the absorption mean free path determines the detector efficiency and the energy resolution. The angular resolution at TeV energies is $\sim 0.7°$ (ice) and $\sim 0.4°$ (water), so the ice and water detectors complement each other.

Also, having IceCube in the Southern hemisphere and KM3NeT (and their prototypes Antares, Nemo and Nestor, as well as Baikal) in the Northern hemisphere provides for a complementary sky coverage by these detectors. This is useful, because neutrinos coming downwards from the atmosphere produce downward muons, which are overwhelmed by downwards secondary muons from cosmic rays which reach the ground. On the other hand, neutrinos coming from the opposite hemisphere (upwards through the detector from below) are barely absorbed by the Earth (at energies $E_\nu \lesssim$ PeV), and if they interact in the detector or close to it, their upwards-moving muons can be measured essentially noise-free in the detector. This is because the muons produced by cosmic rays in the opposite hemisphere are all absorbed in the Earth – only the neutrinos come through. Of course the upward muons will still contain,

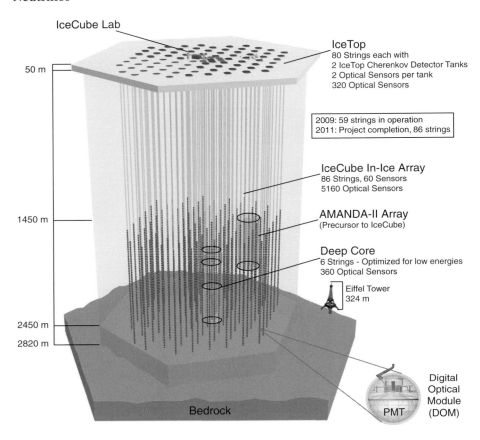

IceCube Lab

50 m

IceTop
80 Strings each with
2 IceTop Cherenkov Detector Tanks
2 Optical Sensors per tank
320 Optical Sensors

2009: 59 strings in operation
2011: Project completion, 86 strings

IceCube In-Ice Array
86 Strings, 60 Sensors
5160 Optical Sensors

AMANDA-II Array
(Precursor to IceCube)

1450 m

Deep Core
6 Strings - Optimized for low energies
360 Optical Sensors

Eiffel Tower
324 m

2450 m

2820 m

Digital
Optical
Module
(DOM)

Bedrock   PMT

Figure 11.7 Schematic diagram of IceCube with phototube strings between
1.4 km and 2.4 km depth in the ice. Also shown are the Eiffel tower for comparison,
and the surface IceTop cosmic-ray detectors.
Source: IceCube collaboration/NSF.

besides the source neutrino muons, also a background of muons produced in
or near the detector by atmospheric neutrinos from the opposite hemisphere,
which are caused by cosmic rays interacting in the opposite hemisphere atmosphere. However, atmospheric neutrinos have a much steeper spectrum than
the typical E_ν^{-2} spectra of astrophysical sources, so the background drops off
with energy, and a good signal-to-noise ratio can be achieved by considering a
small error circle around the source (if the direction is known) and/or a small
time window (if the source is flaring), or by detecting multiple neutrinos in
near-time coincidence from the same direction in the sky.

The Earth becomes opaque for electron and muon neutrinos of energies $E_\nu \gtrsim$
PeV, and one is able to measure these only for angles close to horizontal, or from

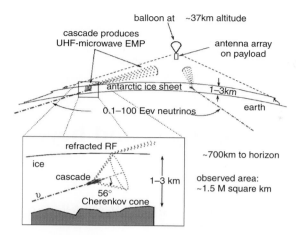

Figure 11.8 ANITA detector concept, with antenna on balloon looking at radio and microwave Cherenkov radiation from grazing incidence EeV neutrinos. *Source*: ANITA collaboration.

above. At these energies, the atmospheric neutrinos and muons are much less, which improves the signal-to-noise ratio. Interestingly, at these energies the Earth remains largely transparent to tau neutrinos, since these "regenerate" by creating tau mesons which again decay into tau neutrinos, in a repeating process which greatly extends the mean free path.

At EeV energies, on the other hand, much larger detectors than IceCube or KM3NeT are desirable. This is because the neutrino spectra are typically $N(E_\nu) \propto E_\nu^{-2}$ or steeper, and there are precious few EeV neutrinos incident on Earth per km^2 per year. To do this, thousands of square kilometers are needed. This can be done with satellites or with balloons which look down at the ice from a large height, with a large field of view downwards. One such experiment, called ANITA, has been flown on a balloon at altitudes of 35–40 km, with anten- nae looking down at the ice (Fig. 11.8). The detection relies on a phenomenon called Askaryan radiation, which is a form of coherent Cherenkov radiation at radio frequencies produced by cascades resulting from grazing incidence EeV neutrinos interacting in the ice. The techniques are being improved, and it is hoped that significant limits may be set on different models of cosmogenic neutrinos.

12

Dark dreams, Higgs and beyond

12.1 Dark matter

As we discussed in Chapter 3, about 26% of the mass-energy density of the Universe at present is in the form of non-relativistic gravitating "matter", and the rest is in some form of "dark energy". Considering for now this 26% non-relativistic matter component, about 1/7th of this, or 4% of the grand total, is in the form of known particles, mostly baryons (the leptons, neutrinos and photons represent much less mass than the baryons). The other 6/7th of the non-relativistic matter, or 22% of the grand total, is something which we call "dark matter", colloquially referred to as DM.

What we know about the dark matter is frustratingly little, but enough to convince us of its existence and to give us an idea of some of its overall properties, as outlined in Chapters 3 and 4. We know that it is there, because its gravity makes itself felt in the dynamics of our galaxy and that of other galaxies, as well as in the dynamics of the expansion of the Universe, and its presence is directly mapped via the "gravitational lensing" effect which distorts the paths of the light rays coming to us from distant objects through foreground dark matter-dominated clusters of galaxies. The DM also plays an important role in determining at what epoch in the expansion of the Universe proto-galaxies start to form and assemble. We also know that it is extremely weakly interacting, if not downright inert. It does not emit, block or reflect light or electromagnetic waves, nor does it seem to interact with other particles, at least not enough to have been detectable so far.

The possibility that the dark matter consists of small black holes, or low mass hard-to-detect dwarfs or planets was considered, and even given the name of MACHOs, or massive compact halo objects (since most of the dark matter in our own galaxy resides in its halo, an ellipsoidal region surrounding our galactic disk and extending beyond it). However, after much searching and probing for these objects, this hypothesis has been shelved. The remaining and most widely embraced view is that it is some form of weakly interacting massive particle (WIMP) – see Chapter 3. It has to be massive in order to be non-relativistic at present, and weakly interacting to explain why we have not detected it so far. And it cannot be neutrinos, since those have a small enough mass to be relativistic at present, and if they made up 22% of the total mass their relativistic motions would have prevented the baryons from assembling into galaxies, as evidently they did. The total mass in neutrinos can be shown to account, at best, for no more than 1 or 2% of the total.

So how can we go about detecting WIMPs? Neutrinos are hard enough to detect, and clearly WIMPs are even harder, since we have not managed to do so despite 20 years of trying. Remarkably, if we assume that WIMPS make up today 22% of the energy density of the Universe, and if we assume that the strength of the interaction (the coupling constant, in particle physics jargon) is comparable to that of the weak interactions, then by considering the thermal evolution of the early Universe when WIMPS ceased to be tightly coupled to the rest of matter, we can make a rough estimate. This results in a range for the WIMP mass of $m_W c^2 \sim 0.1 - 1$ TeV, corresponding to a WIMP cross-section for interacting with normal matter of $\sigma_W \sim 10^{-36}$ cm^2. This cross-section is very small, but it is comparable to that of TeV neutrinos being measured with IceCube, and it is also within the reach of large accelerators such as the Large Hadron Collider in Geneva, and other planned experiments. WIMPs are not part of the Standard Model of particle physics, and for this reason they are of extreme interest. But, by the same token, the details of their interactions are rather model-dependent, and there is a variety of models between which to decide.

There are two main approaches in the search for dark matter. One is called the indirect approach, which consists of searching for astrophysical signals resulting from the weak interactions of WIMPs in the Universe. The other is called the direct approach, consisting of looking for signals from the weak interactions of WIMPs with normal matter in specially designed laboratory experiments.

12.2   Indirect astrophysical WIMP searches

The specific type of interactions and the secondary particles of WIMPs interacting with normal matter depend on the specific BSM (beyond the

Standard Model) theory used as a guideline. The most widely used family of BSM theories involve some form of supersymmetry (see Chapter 2), in which the favored WIMP particle is called a *neutralino*, symbolized as χ. In these models the χ and their anti-particles $\bar{\chi}$ can annihilate, resulting in (ultimately) Standard Model pairs of particles, such as neutrinos and anti-neutrinos, electrons and positrons, photons, etc.,

$$\chi + \bar{\chi} \rightarrow \nu + \bar{\nu}$$
$$+ e^+ + e^-$$
$$+ \gamma + \gamma \cdots . \tag{12.1}$$

Being very massive and non-relativistic, WIMPs are expected to populate not only the halo of our galaxy and clusters of galaxies, but they might also have accumulated in our galactic center and at the center of the Earth or the Sun. The annihilation rate can depend not only on the mean density of WIMPs, but also on a possible clumpiness of the DM. In the case of the Sun or the Earth, decay products such as e^+e^- or photons would be absorbed, so we can only look for annihilation neutrinos (e.g., with IceCube, or with Km3NeT).

Recent limits from the IceCube 22-string detector were obtained based on the number of muon events attributed to muon neutrinos coming from the direction of the Sun compared to other directions. These results, assumed to be from WIMP annihilation neutrinos, can be translated into a limit for a spin-dependent neutralino-proton cross-section versus mass, which is to be compared with some of the direct experimental limits, shown in Fig. 12.1.

On the other hand, for annihilation of WIMPs in the galactic halo or the galactic center one would expect to see gamma-rays (e.g., with Fermi, VERITAS, etc.) as well as charged particles (e.g., with PAMELA, ATIC, etc.). For instance, the PAMELA satellite announced [70] an excess of $\sim 10\%$ in the flux of positrons above what is expected from cosmic-ray interactions in the interstellar medium, see Fig. 12.2. The excess signal has been interpreted as being due to DM annihilations. However, this conclusion is highly tentative, as to be correct it would require an additional "boost factor" of $\sim 10^4$ in the annihilation rate over what is expected from the DM energy density in the halo, which is not easily accounted for by invoking clumpiness. Astrophysical explanations, such as positrons from nearby pulsars (e.g., Geminga) [71], nearby supernovae or simply cosmic-ray propagation effects could also plausibly explain the excess. An even larger excess above the usual predicted cosmic-ray secondary flux up to 800 GeV was announced by the ATIC experiment, which was widely suspected of being the signal of DM annihilation. However, the Fermi spacecraft subsequently published results [72] on the cosmic-ray electron spectrum up

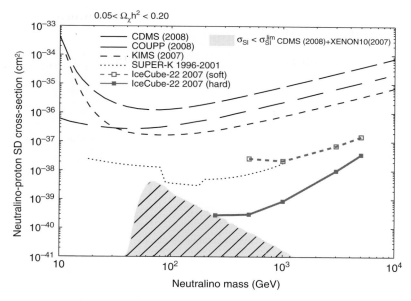

Figure 12.1 IceCube 22-string upper limits for the 90% confidence level on the spin-dependent neutralino-proton cross-section for soft (lower dashed line) and hard (lower full line) annihilation channels, as a function of neutralino mass. The limits from some other experiments are also shown. The diagonally hatched area represents minimal super-symmetric models not disfavored by direct searches based on spin-independent limits from various experiments. From [69].

Souce: Copyright 2009 by The American Physical Society.

to $\sim$ TeV (Fig. 12.2), showing only mild departures from the conventional cosmic-ray secondary predictions, which can be fitted with minimal adjustments to the latter.

12.3 Direct WIMP searches

The *direct* detection of WIMPs relies mainly on measuring the nuclear recoil of nucleons impacted by a WIMP in the lab. The cross-section depends on the WIMP model, typically based on super-symmetric theories. Depending on the model and on the detector, in some cases the relevant cross-section is that for spin-independent scattering between WIMPs and nucleons, in which the scattering amplitudes from the various nucleons (protons and neutrons) in the impacted target nuclei add up together coherently. This amplifies the net cross-section by a factor proportional to A^2, where A is the atomic number of the detector material (i.e., the total number of protons and neutrons in the target nucleus). In other cases, the relevant cross-section is that for spin-dependent

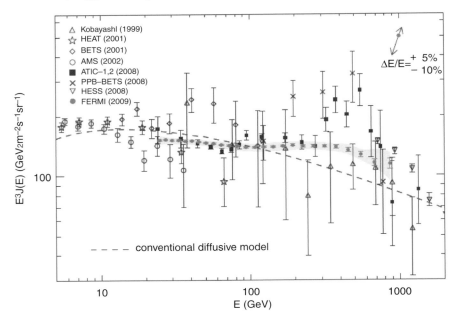

Figure 12.2   Fermi observations of the cosmic-ray electron spectrum, compared to various other experiments, including ATIC [72].
Source: NASA/DOE/Fermi LAT Collaboration. Copyright 2009 by The America Physical Society.

scattering. In this case the scattering amplitudes add up incoherently from the different nucleons, because most of the nucleon spins cancel out. This cross-section is therefore a factor A^2 smaller than that for coherent scattering.

The event rate (the number of recorded WIMP hits) depends on the cross-section as well as on the incident WIMP flux. Because of their mass, WIMPs are expected to be gravitationally concentrated in the galaxy halo and towards the galactic center, so the density in the solar neighborhood is estimated to be $\sim 10^5$ larger than the universal average. For typical WIMP masses $m_\chi \sim 100\,\mathrm{GeV}$ and $A \sim 70$, as an example, this might imply rates $R \sim 5$ per day per kg of detector mass for coherent (spin-independent) scattering, and $R \sim 10^{-3}$ events per day per kg of detector mass for incoherent (spin-dependent) scattering. These are only indicative values, since they depend on many parameters of the SUSY or other models, including also the WIMP coupling constant.

Attempts to detect the recoiling nuclei rely on different methods, such as measuring the ionization trail, or measuring the lattice vibrations (phonons excited) by energy deposition of the recoiling nuclei. One way of distinguishing such events from the enormous background due to radioactive decays, cosmic rays, etc., is to place the experiments deep underground. Another discriminant is that the WIMP event rate is expected to be modulated, with an angular and time dependence introduced by the motion of the Earth and the Sun. For example, if WIMPs populate predominantly the halo of the galaxy, there could be a daily shadowing effect by the Earth when it is turned away from the galactic center. One can also expect an annual modulation of the event rate, since the WIMP velocity distribution and the cross-section change as the velocity of the Earth around the Sun ($\sim 30\,\mathrm{km\,s^{-1}}$) is added or subtracted from the velocity of the Sun around the galactic center ($\sim 230\,\mathrm{km\,s^{-1}}$), resulting in annual event rate changes on the order of 5% [73]. The DAMA/LIBRA experiment group has in fact reported such an effect [74]. These results have not been confirmed by other experiments using different techniques, stimulating the formulation of possible theoretical dark matter models aimed at simultaneously satisfying the DAMA, PAMELA, ATIC and Fermi results (e.g., [75]). Large detector masses of extremely pure detector material located deep underground are advantageous. Some experimental limits are shown in Fig. 12.3. The detectors discussed here are all in

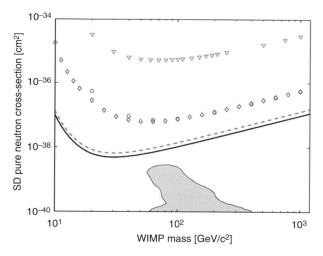

Figure 12.3   Spin-dependent WIMP nucleon cross-section direct search limits as a function of WIMP mass, probed by the XENON10 experiment for pure neutron coupling (solid line) and for an alternative model (dashed line). Also shown are the experimental results from the CDMS experiment (diamonds), ZEPLIN-II (circles), and KIMS (triangles). The theoretical filled region below the limit curves is what is expected for the neutralino in the constrained minimal super-symmetric model [76].

Source: Copyright 2008 by The American Physical Society.

the process of being upgraded to larger detector masses (e.g., Super-CDMS is projected to reach 25 kg, and XENON1T is projected to reach a 1 ton-year exposure).

12.4   Axions

Another dark matter particle candidate is the *axion*, a hypothetical light boson, which was introduced theoretically to explain a problem in strong interaction physics, namely the fact that strong interactions do not seem to violate a symmetry called charge-parity symmetry, something which would have been expected at some level. This was solved by Peccei and Quinn with a theory which required introducing a new light particle called the axion (e.g., [1, 77]). Its mass m_a and its very small coupling to other particles depends on a single parameter, the unknown (Peccei–Quinn) energy scale, estimated to be $f_a \sim 10^9$–10^{13} GeV, leading to a weak coupling and an axion mass $m_a = (6 \times 10^{15} \text{ eV}/f_a) \sim 10^{-6}$–$10^{-2}$ eV. This axion mass range, which is of cosmological interest (i.e., leading to $\Omega \sim 1$) is currently not easily detectable. However, it is weakly interacting, and non-relativistic, having formed as a condensate in the early Universe, and this makes it a possible candidate for dark matter.

The axion, being neutral, can decay into two photons, but the lifetime for this is extremely slow. This rare annihilation into two photons, each of energy $m_a c^2/2$, would have shown up as a narrow line in optical telescopes if the mass m_a were in the 0.1 to few eV range, which can thus be ruled out. Also, masses above a few eV can be ruled out, since they would have led to a drastic reduction in the number of He-burning red giants. They would lose energy by axion emission, which would have to be compensated by increased nuclear burning, shortening the lifetime of the stars and their observed abundance. The most promising efforts rely on another property of axions, which is that it can also interact with strong magnetic fields, leading to conversion of an axion into a real photon. Measurements in pulsars appear unrealistic, due to the low fluxes, but efforts to observe this effect in microwave cavities in the lab are underway. A different type of experiment at CERN, called CAST (CERN Axion Solar Telescope), looks at the Sun using a 9.26 m long LHC magnet with fields up to 9.5 teslas to convert axions into X-rays, which are continuously recorded. This experiment is still ongoing.

12.5   Dark energy

We discussed in Chapter 3 that at present, and in fact in the recent past as well, between redshifts $0.5 \lesssim z \lesssim 1$, the expansion of the Universe appears to have been accelerating. This leads us to infer that the currently dominant form of mass-energy density in the Universe is some form of vacuum energy. This can be represented through the famous Einstein cosmological constant Λ, so one can write formally the vacuum (or dark) energy density as

$$\rho_V = \frac{\Lambda}{8\pi G}. \tag{12.2}$$

In order to result in the observable accelerating expansion, the "equation of state" of this dark energy, that is a relation between its energy density $\rho_V c^2$ and its pressure p, must be of the form $p = w\rho_V c^2$ with $w = $ a constant, at least over a limited range of redshift, and what is strange, w must be negative (i.e., it must have a *negative* pressure). This implies a negative gravitational attraction in the Friedmann equations (Einstein's equations written for a homogeneous isotropically expanding universe); that is, it implies an acceleration of the expansion. In fact, cosmological measurements from the WMAP cosmic background experiment, combined with high redshift supernova surveys, show that to a very good approximation the constant $w \simeq -1$, and hence the equation of state of the vacuum energy, is close to $p \simeq -\rho c^2$, which in particle physics and solid state physics is encountered for a type of field called a scalar field (i.e. fields which are represented by only one number at each position in space-time, such as the normal mass density; this is in contrast to fields like the electromagnetic fields, which require three quantities to describe them at each point in space-time – since they are vector fields, they have not only a strength but also a direction).

Using the definition of the critical energy density (3.2) in Chapter 3, the cosmological total density parameter today is $\Omega_0 = (\rho_{tot,0}/\rho_{c,0}) \simeq 1$; that is, the Universe appears to have the "critical" density $\rho_{c,0} = 3H_0^2/8\pi G = 1.88 \times 10^{-29}h^2 \,\mathrm{g\,cm^{-3}} \simeq 5.1h_{70}^2 \,\mathrm{keV\,cm^{-3}}$, where H_0 is the Hubble constant characterizing the observed expansion velocities, $H_0 \simeq 100h \,\mathrm{km\,s^{-1}\,Mpc^{-1}}$, and measurements favor a value of $H_0 = 70h_{70} \,\mathrm{km\,s^{-1}\,Mpc^{-1}}$. If we separate the contributions to Ω_0 from non-relativistic dark matter (M) and from dark energy (V), the former amounts to 26% and the latter amounts to 74%, or

$$\rho_V \sim 6.3 \times 10^{-9} \,\mathrm{erg\,cm^{-3}} \simeq 3.9 \times 10^3 \,\mathrm{eV\,cm^{-3}} \sim 4\,\mathrm{keV\,cm^{-3}}. \tag{12.3}$$

This is equivalent to one medium energy X-ray photon per cubic centimeter. Another comparison is to consider an electric field of $1\,\mathrm{V\,m^{-1}} = 3 \times 10^{-5} \,\mathrm{statvolt\,cm^{-1}}$, which is easily measurable in the lab. Thus, a direct measurement of the dark energy might appear in principle possible. However, the problem is that one does not know what kind of field is involved in the dark energy.

A striking property of the present inferred dark energy density ρ_V is that it would be negligible at early times ($z \gtrsim$ few), but it eventually comes to dominate all forms of matter (and curvature) energy as we approach the present epochs, at $z \to 1$. It is a puzzle why $\Omega_V \sim 1$ precisely now. This implies a density close to critical, $\rho_V \sim 10^{-29}\,\mathrm{g\,cm^{-3}}$. From quantum field theory, one would naively associate with such a vacuum energy density some quanta, or particles, of mass $m_V \sim 10^{-11.5}\,\mathrm{GeV} \sim 10^{-2.5}\,\mathrm{eV}$, which would have been expected to be some fundamental mass scale in nature. However, the natural mass scale for gravity is the Planck mass, $m_{Pl} = 1.2 \times 10^{19}\,\mathrm{GeV} \sim 10^{19}m_p \sim 10^{-5}\,\mathrm{g}$, which is enormously greater, and with which one would associate an energy density which is 122 orders of magnitude larger (i.e., one would have expected $\Omega_V \sim 10^{122}$, but instead we have $\Omega_V \sim 1$!). If it is so small, why not zero? Another puzzle is that $\Omega_V \sim \Omega_M \sim 1$ at present, but since $\Lambda = $ constant, ρ_V must have been essentially the same in the past, whereas we know that $\rho_M \propto (1+z)^3$ must have increased rapidly as we go back into the past. So why did they become comparable precisely now, and not earlier or later? These are basic questions whose answers may depend on a theory of quantum gravity.

Thus, dark energy appears to be far more difficult to tackle than dark matter. The latter at least has a richness of plausible interaction models, which, in addition to its gravitational properties, lead to predictions about various types of particle interactions that can be searched for, directly or indirectly. In the case of dark energy, there are no compelling theoretical models, and it could well be described as a massive headache. One is left with the (well-observed and fairly well-established) gravitational effects, and these are the only type of (indirect) measurements that are at present being actively pursued. This includes large-scale projects in various stages of preparation, to survey from low to medium redshifts the rate of expansion and the rate of assembly of large-scale structures (since the assembly of structures is hindered by the growing acceleration). The techniques include the mapping of source distributions with redshift using direct distance measurements with supernova Ia, distant galaxies, quasars or gamma-ray bursts, mapping mass distributions with weak gravitational lensing, and a number of other approaches.

12.6   Beyond the Standard Model at the LHC

The LHC at the European Center for Nuclear Research (CERN) in Geneva, Switzerland started limited operations in 2009. Its primary stated goals include to look for the last missing link in the Standard Model of particle physics, the famous Higgs boson, which is a key piece in our current thinking about the Universe – both the early Universe and the lab universe. Beyond that, it aims to explore interactions beyond the Standard Model, and this includes, among others, looking for dark matter.

The LHC has 1232 superconducting magnetic dipoles, each ~ 14.3 m long, operating at a temperature $T = 1.9$ K (that is, $-456.25°$ F, or $-271.25°$ C – very close to absolute zero!). These magnets bend the protons to go along a circular underground tunnel of radius $R = 27$ km under meadows and villages in the Swiss and French Alps. At a proton energy $E_p = 7$ TeV per beam, the magnets need to produce a magnetic field of $B \simeq 8.33$ tesla = 8.33×10^4 gauss, requiring a current of $I = 1.17 \times 10^4$ amps. The LHC dipole coils have 7600 km of superconducting cables whose total weight is 1200 tons [78].

The principle of colliders such as the LHC is to have charged particles revolve in opposite directions, so that they eventually meet head-on. The center of momentum energy of the counter-rotating protons is 14 TeV. The frequency of revolution is $\Omega_{rev} = c/27$ km $\sim 10^4$ Hz, and the instantaneous *luminosity* is $\mathcal{L} \sim 10^{34}$ cm^{-2} s^{-1}. For a proton–proton cross-section $\sigma_{pp} \sim 6 \times 10^{-25}$ cm^2, the collision rate is $\mathcal{L} \times \sigma \simeq 10^9$ Hz; that is, 10^9 (a billion) collisions per second. This huge number of collisions is what makes possible looking for the extremely rare events that would provide information about the Higgs particle, dark matter or BSM physics. The *total* energy in the beam is $E_b \simeq 0.185 M_{Pl}c^2 = 364$ MJ, which is equivalent to the joint kinetic energy of 100 medium-sized pick-up trucks ($M \sim 3$ tons each) going at 100 miles (160 km) per hour.

The main detectors in the LHC are called ATLAS (A Toroidal LHC ApparatuS), and CMS (Compact Muon Solenoid) (see Fig. 12.4). These will look for the Higgs

Figure 12.4 The ATLAS detector on the LHC at CERN, Geneva, showing the eight toroidal magnets surrounding the calorimeters before moving the latter into the middle of the magnets. The scientist at the bottom provides a scale comparison. *Source:* CERN.

particle, whose mass is estimated to be in the range of the so-called electroweak energy scale, or $m_H \sim 180\,\text{GeV}$. They will also look for decay particles signaling interactions with, or resulting from, the decay of super-symmetric particles inhabiting the BSM realm. It is also hoped that one or more of these super-symmetric particles may be identified as providing the dark matter WIMPs implied by cosmology.

Higgs searches are a high priority for the LHC and other accelerators. A difficulty is that the Higgs mass is unknown from theory, only the strength of its coupling to various particles, which is proportional to the mass of the particles. Thus, it is expected to couple lightly to neutrinos, leptons and light quarks such as the u, d and s, and much more strongly to heavy particles such as the $W^{\pm}$, Z^0 and the t quark. Thus, the first difficulty is to produce in sufficient numbers the heavy particles with which it interacts. The Higgs boson must then be detected via its decay product against a large background arising from other strong interaction processes. For Higgs masses less than twice the W-boson mass, the Higgs bosons will decay into fermion–anti-fermion pairs, but this signal however is swamped by other hadronic events. Thus, in this energy range plans are to rely on the much rarer but distinctive decay mode into two photons, $H^0 \rightarrow \gamma + \gamma$. This decay mode has a branching ratio (relative probability) of only 10^{-3}, so data accumulation would be slower, but the background is lower. On the other hand, if the mass of the Higgs exceeds twice the Z-boson mass, the dominant decay modes are into pairs of W bosons or pairs of Z bosons, $H^0 \rightarrow Z^0 + Z^0$ and $H^0 \rightarrow W^+ + W^-$. The W and Z bosons must then be identified via their own decay products. The most distinctive signature occurs when both Z^0 decay into electron or muon pairs,

$$H^0 \rightarrow Z^0 + Z^0 \rightarrow \ell^+ + \ell^- + \ell^+ + \ell^-, \tag{12.4}$$

where the $\ell^{\pm} = (\nu^{\pm}, e^{\pm})$ are the leptons being searched for. Although only 4% of the decays result in this signature, it is referred to as the "gold-plated" decay channel, because it is so distinctive that it would allow Higgs bosons to be detected in the range $200\,\text{GeV} \leq M_H \leq 500\,\text{GeV}$ soon after the LHC becomes operational. A hypothetical Higgs event at the LHC is illustrated in Fig. 12.5.

WIMPs are weakly interacting, and hence do not carry electric or color charge, but they are expected to originate from heavier parent particles which need not be weakly interacting, and hence would be unstable to decay into the lightest stable WIMPs (the actual dark matter) plus Standard Model particles. These energetic WIMP-related Standard Model particles can be detected by the LHC, but when looking at the dynamics of the interaction there will be what is called "missing transverse energy", due to WIMPs which escape the detector

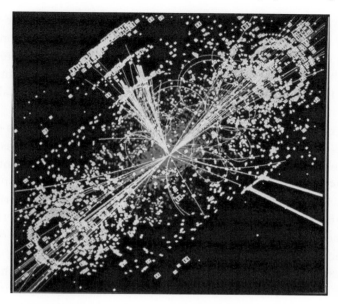

Figure 12.5 A simulated event showing the decay of a Higgs particle following the collision of two protons in the CMS experiment at the LHC in the European particle physics institute, CERN. A detection of the Higgs particle is one of the last missing pieces in the Standard Model, and it is a major goal of the LHC. *Source*: CMS team, CERN.

without leaving a trace of themselves (other than the fact that their energy and momentum is "unaccounted" for). In *pp* collisions in the Standard Model, neither the total energy nor the total longitudinal momentum of a collision can be predicted accurately, but the total transverse momentum can be predicted. This momentum is measurable by the LHC, which allows us to establish any missing transverse momentum and energy. Of course, other neutral particles such as Standard Model neutrinos also produce missing transverse energy signals (see Fig. 2.2), which act as a background that must be accounted for and subtracted. After that is done, the remaining missing transverse energy is what can signal the presence of WIMPs.

12.7   Underground astrophysics and particle physics

It would seem insane to pursue astrophysical observations from deep underground sites, yet such experiments have already yielded immensely valuable information about solar neutrinos and neutrino physics, through observations made in the Homestake mine (1.5 km depth), Kamiokande (1 km depth) and the Sudbury Neutrino Observatory, SNO (2.07 km depth). The reason for this is that neutrinos are detected through secondary charged leptons such as muons and e^+e^- which they produce in the detector, but if the detector is not shielded there would be a huge background of secondary muons and e^+e^- from cosmic rays which would overwhelm the neutrino secondary signal being searched for.

In fact there are a number of other particle physics or astrophysics experiments, such as the search for proton decay, neutrino oscillations or direct dark matter searches, which also require a very "cosmic-ray-free" environment, since they involve very weak signals which could easily be masked by cosmic-ray-induced noise. The Earth provides a good filter, and the deeper the overburden or the path length through the filter so much the better. This is also the reason why TeV neutrino experiments such as IceCube or KM3NeT aim to observe neutrinos which have traversed the Earth on their way to the detector.

The implications of these experiments for probing physics beyond the Standard Model of particle physics are far-reaching. Proton decay is predicted by grand unified theories and its measurement would probe fundamental interactions in the 10^{16} GeV energy range (compared to the 10^4 GeV = 10^{13} eV range of the LHC!). Dark matter direct detection would not only be of major impact for astrophysics, but would probe ideas about super-symmetric grand unified theories, which posit an equivalence between fermions and leptons. Neutrino oscillation experiments would probe charge–parity symmetry violation, which could provide information about why the Universe is dominated by matter rather than anti-matter.

The underground laboratories at Kamioka (Japan), Sudbury Neutrino Observatory (Canada), Gran Sasso (Italy), and Boulby (UK) have been conducting various experiments for some time now, and have undergone continuous upgrades of their facilities. Among these are experiments on non-baryonic dark matter, which are reaching a sensitivity that brings them close to being able to detect neutralinos under more or less optimistic theoretical expectations. Other experiments continue the long search for proton decay, initiated almost 30 years ago (which, while continuing their original primary purpose, yielded serendipitously the discovery of supernova neutrinos from SN 1987a). Another set of experiments are probing the next phase of neutrino physics on solar and supernova neutrinos and on atmospheric neutrinos. Gran Sasso is also conducting experiments to probe whether neutrinos and anti-neutrinos are one and the same; that is, whether they are their own anti-particle (this is referred to as the Majorana versus the Dirac nature of electron neutrinos), which is being explored with the search for the so-called neutrinoless double-beta decays of certain isotopes. Soudan, Gran Sasso and Kamioka are also pursuing complementary sets

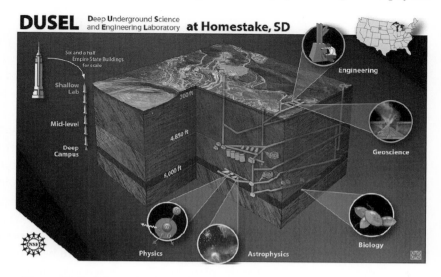

Figure 12.6   Schematic of experiments in the DUSEL underground laboratory planned at Homestake, South Dakota.
Source: Zina Deretsky, National Science Foundation.

of experiments on long base-line neutrino oscillations using the muon–neutrino beams from Fermilab, CERN, KEK, and Tokai.

A major new Deep Underground Science and Engineering Laboratory (DUSEL) has been approved in the USA, which will be located deeper (down to 2.25 km) in the Homestake mine of South Dakota (see Fig. 12.6). This will consist of a very large megaton water Cherenkov detector for astrophysical neutrino and nucleon decay detection, which is roughly 20 times the size of current comparable detectors [79]. It also plans to house experiments for dark matter detection, neutrinoless double-beta decay, to operate long base-line measurements in combination with the FermiLab accelerator neutrino beam, and to conduct experiments aimed at probing the existence of neutron–anti-neutron ($n \leftrightarrow \bar{n}$) oscillations. The latter seek to answer whether neutrons are their own anti-particle, a question which would have far-reaching implications for understanding the origin of matter in the Universe, complementing the studies of proton decay. These probe the ultimate stability of matter, testing the so-called conservation of baryon number, which is one of the fundamental principles on which our current theories of particle physics are based. Ultimately, they seek to answer the question "where did we come from"? This is an ambitious goal, which has secondary ramifications for understanding the nature of dark matter, as well as having implications for more exotic theories of quantum gravity.

Epilogue

The Universe had a fiery beginning, aptly described by the name of Big Bang, involving energies and densities which are thought to be so extreme as to require risky theoretical extrapolations of our limited empirical physical knowledge. This is especially true for the initial period extending from times of about 10^{-44} s down to about 10^{-10} s, when the energies involved in particle collisions exceeded those currently achievable in particle accelerators such as the LHC. Even for epochs $t \gtrsim 10^{-5}$ s, where the expected energies have been explored in the lab, the densities are such that one expects individual quarks and gluons to exist in a plasma phase, where they have not yet combined into hadrons such as protons, neutrons and mesons. This is a regime which has only recently started to be explored with the Relativistic Heavy Ion Collider (RHIC) at Brookhaven.

Beyond this hadronization epoch, however, the Universe entered a period where we are fairly confident that we understand the physics well enough to be able to trace its evolution with reasonable accuracy. As it continued to cool and became increasingly less dense, aside from an episode of fairly well-understood nucleosynthesis between 1 and about 100 s, it would seem as if the Universe had embarked on a peaceful journey of increasing placidity and simplicity. Normal and dark matter coexisted without interfering with each other, and dark energy played an apparently negligible role during this epoch. The tiny perturbations in the Universe grew very slowly, without appreciably disturbing the apparent smoothness until fairly recent epochs. The fires had, it appeared, slowly calmed down.

This simple and accurately quantifiable picture, however, started to change around the time the Universe approached one billion years of age after the Big Bang. The density perturbations reached "non-linear" values which were comparable to the average value, and started to collapse into much denser self-gravitating bodies, the future stars, galaxies and clusters of galaxies. Our understanding of these "late" phases of the Universe is much less exact, due to the increased complexity of the gas dynamic and many-body phenomena involved. Turbulence, instabilities and chaotic motions developed, whose understanding both in and out of the laboratory is hard to predict, just as it is hard to predict oceanic wave patterns or the weather. We have to rely on phenomenological fits with adjustable model parameters, and these provide a reasonably good overall understanding, even if it does not extend to predicting the details accurately.

During this more recent epoch the cores of the more massive primeval stars collapsed into black holes, which became increasingly abundant in the central regions of the larger proto-galaxies, where central black holes grew through mergers and accretion of the surrounding gas and stars. This rekindled the dormant furnaces in the Universe, which now reappeared highly concentrated in the regions around the newly formed black holes. This resulted in local focal points of enormously higher densities and temperatures than in the surrounding average medium, accompanied by violent large-scale plasma motions leading to magnetic fields, charged particle acceleration, and high to ultra-high energy non-thermal radiation.

The particles and field energy densities in these compact focal regions mirror, in fact, the more widely prevalent conditions in the early Universe before the hadronization era, during the epochs where our understanding is most fragmentary and speculative. The fiery monster was only slumbering, and like a many-headed hydra it has reappeared in a multiplicity of violent high energy hot-spots. Unlike Heracles, however, we do not need to tackle these monsters, what we want is to understand them.

We are incredibly lucky in that the TeV energy neutrinos being observed now are typical of energies expected in the Universe around the electroweak era around 10^{-10} s, and the 10^{20} eV energy cosmic rays being observed now have energies characterizing the Universe at around 10^{-28} s, not much after the inferred epoch of inflation. Astrophysical black holes provide laboratories for probing strong fields and quantum gravity ideas which would have been dominant in the early Universe. We can be optimistic that particle astrophysics experiments and theoretical work on high energy astrophysical sources, combined with laboratory and accelerator experiments, will lead to fruitful advances in our understanding both of our present earthly and cosmic surroundings and of our cosmic history.

References

[1] Wilczek, F. (2008) *The Lightness of Being*. New York: Basic Books.

[2] Randall, L. (2005) *Warped Passages*. New York: Harper Collins.

[3] http://dsc.discovery.com/news/2008/09/10/black-hole-cern.html

[4] Giddings, S. B. and Mangano, M. L. (2008) *Phys. Rev. D*, 78.035009.

[5] Greene, B. (2003) *The Elegant Universe: Superstrings, hidden dimensions and the quest for the ultimate theory*. New York: W.W. Norton.

[6] Smolin, L. (2001) *Three Roads to Quantum Gravity*. New York: Basic Books.

[7] Oriti, D. (ed.) (2009) *Approaches to Quantum Gravity*. Cambridge: Cambridge University Press.

[8] Rees, M. J. (2003) *Our Cosmic Habitat*. Princeton, NJ: Princeton University Press.

[9] Coughlan, G. D., Dodd, J. E. and Gripaios, B. M. (2006) *The Ideas of Particle Physics*. Cambridge: Cambridge University Press. 3rd edn.

[10] Solomey, N. (1997) *The Elusive Neutrino*. Scientific American Library; distributed by W. H. Freeman and Co., New York.

[11] Perkins, D. H. (2003) *Particle Astrophysics*. Oxford: Oxford University Press.

[12] Martin, B. R. and Shaw, G. (1997) *Particle Physics*. Chichester: John Wiley & Sons.

[13] Thorne, K. S. (1997) *Black Holes and Time Warps*. New York: Norton.

[14] t'Hooft, G. (1997) *In Search of the Ultimate Building Blocks*. Cambridge: Cambridge University Press.

[15] Coles, P. (2001) *Cosmology: A Very Short Introduction*. Oxford: Oxford University Press.

[16] Longair, M. (1998) *Galaxy Formation*. Berlin: Springer.

[17] Komatsu, E. *et al.* (2009) *ApJS* **180**: 330.

[18] Silk, J. (2008) *The Infinite Cosmos*. Oxford: Oxford University Press.

[19] Peacock, J. A. (1999) *Physical Cosmology*. Cambridge: Cambridge University Press.

[20] Rees, M. J. (2000) *New Perspectives in Astrophysical Cosmology*. Cambridge: Cambridge University Press.

[21] Springel, V. *et al.* (2005) *Nature* **435**: 629.

[22] Rees, M. J. and Ostriker, J. P. (1977) *MNRAS* **179**: 541.

[23] Silk, J. (1977) *ApJ* **211**: 638.

[24] Garwin, R. L. and Charpak, G. (2001) *Megawatts and Megatons*, p. 210. New York: Alfred Knopf.

[25] Hartle, J. B. (2003) *Gravity*. San Francisco: Addison Wesley.

[26] Begelman, M. and Rees, M. (2009) *Gravity's Fatal Attraction*. Cambridge: Cambridge University Press. 2nd edn.

[27] Melia, F. (2003) *The Edge of Infinity*. Cambridge: Cambridge University Press.

[28] Brandt, W. N. and Hasinger, G. (2005) *Ann. Rev. Astron. Astrophys.* **43**: 827.

[29] Genzel, R. and Karas, V. (2007) In *Black Holes from Stars to Galaxies – Across the Range of Masses*, eds V. Karas and G. Matt. Proc. IAU Symp. No. 238, p. 173 (arXiv:0704.1281).

[30] Do, T. *et al.* (2009) *ApJ* **691**: 1021.

[31] Reid, M. J. *et al.* (2009) *ApJ* **695**: 287.

[32] Reid, M. J. *et al.* (2009) *ApJ* **695**: 287.

[33] Peterson, B. and Horne, K. (2006) In *Planets to Cosmology*, eds M. Livio and S. Casertano (Space Telescope Science Institute Symposium Series), p. 124 (arXiv:0407538).

[34] Krolik, J. H. (1999) *Active Galactic Nuclei*. Princeton, NJ: Princeton University Press.

[35] Rybicki, G. and Lightman, A. (1979) *Radiation Processes in Astrophysics*. New York: John Wiley & Sons.

[36] Wheeler, J. C. (2007) *Cosmic Catastrophes*. Cambridge: Cambridge University Press.

[37] Soderberg, A. M. *et al.* (2008) *Nature* **453**: 469.

[38] Aharonian, F. *et al.* (2009) *ApJ* **696**: L155.

[39] Weisberg, J. M. and Taylor, J. H. (2004) (arXiv:astro-ph/0407149); also Hulse, R. A. and Taylor, J. H. (1975) *ApJ*, **195**: L51.

[40] Gehrels, N., Ramirez-Ruiz, E. and Fox, D. B. (2009) *Ann. Rev. Astron. Astrophys.* **47**: 567 (arXiv:0909.1531).

[41] Abdo, A. A. and the Fermi collaboration (2009) *Science* **323**: 1688.

[42] Mészáros, P. (2006) *Rep. Prog. Phys.* **69**: 2259.

[43] Abdo, A. A. and the Fermi collaboration (2009) *Nature* **462**(7271): 331.

[44] Abdo, A. A. *et al.* (2009) *Nature* **462**: 7271, p. 331–334.

[45] Abdo, A. A. and the Fermi collaboration (2009) *ApJ* **699**: L102.

[46] Abdo, A. A. and the Fermi collaboration (2009) *ApJ* **700**: 1059.

[47] Khangulyan, D., Hnatic, S., Aharonian, F. and Bogovalov, S. (2007) *MNRAS* **380**: 320.

[48] Weekes, T. *et al.* (1989) *ApJ* **342**: 379.

[49] Reynolds, S. P. (2008) *Ann. Rev. Astron. Astrophys.* **46**: 89.

[50] Aharonian, F. *et al.* (2009) *ApJ* **696**: L155.

[51] Sikora, M., Begelman, M. and Rees, M. J. (1994) *ApJ* **421**: 153.

[52] Coppi, P. and Aharonian, F. (1997) *ApJ(Lett.)* **487**: L9.

[53] Hulse, R. A. and Taylor, J. H. (1975) *ApJ* **195**: L51.

[54] O'Shaughnessy, P., Kalogera, V. and Belczynski, T. (2009) arXiv:0908.3635.

[55] Flanagan, E. E. and Hughes, S. A. (2005) *NJPh* **7**: 204.

[56] Hobbs, G. *et al.* (2009) In Proc. Amaldi 8 Conference (also: arXiv:0907.4847).

[57] Cronin, J. W. (2004) arXiv:astro-ph/0402487.

[58] Abraham, J. and the Pierre Auger Collaboration (2008) *Phys. Rev. Lett.* **101**: 061101.

[59] Olinto, A. V. *et al.* arXiv:0903.0205 [astro-ph].

[60] The Pierre Auger Collaboration *et al.* (2007) *Science* **318**: 938.

[61] Waxman, E. (2006) *AIPC* **836**: 589.

[62] http://www.nature.com/physics/highlights/6907-4.html

[63] Bahcall, J. N. (1989) *Neutrino Astrophysics*. Cambridge: Cambridge University Press.

[64] Gaisser, T. K. (1990) *Cosmic Rays and Particle Physics*. Cambridge: Cambridge University Press.

[65] Christiansen, H., Romero, G. and Orellana, M. (2007) *Braz. J. Phys.* **37**: 545.

[66] Waxman, E. and Bahcall, J. N. (1999) *PRD* **59**: 023002.

[67] Stanev, T. (2004) *High Energy Cosmic Rays*. Berlin: Springer.

[68] Halzen, F. (2009) *J. Phys. Conf. Ser.* **171**: 2014.

[69] Abbasi, R. and the IceCube collaboration (2009) arXiv:0902.2460.

[70] Adriani, O. *et al.* [PAMELA collaboration] (2008) arXiv:0810.4995 [astro-ph].

[71] Morselli, A. and Moskalenko, I. V. (2008) arXiv:0811.3526 [astro-ph].

[72] Abdo, A. A. and the Fermi collaboration (2009) *Phys. Rev. Lett.* **102**: 1101.

[73] Hooper, D. and Baltz, E. A. (2008) arXiv:0802.0702 [hep-ph].

[74] Bernabei, R. *et al.* (2008) *Eur. Phys. J.* **C56**: 333 (arXiv:0804.2741).

[75] Arkani-Hamed, N. *et al.* (2009) (arXiv:0810.0713).

[76] Angle, J. *et al.* (2008) *Phys. Rev. Lett.* **101**: 091301.

[77] Kolb, R. W. and Turner, M. S. (1990) *The Early Universe*. Reading, MA: Addison-Wesley.

[78] Kane, G. and Pierce, A. (eds) (2008) *Perspectives in LHC Physics*. Singapore: World Scientific.

[79] Raby, S. *et al.* (2008) DUSEL Theory White Paper, arXiv:0810.4551 [hep-ph].

Glossary

$L_\odot$	the luminosity of the Sun – $L_\odot \simeq 4 \times 10^{33}\,\mathrm{erg\,s^{-1}}$, 56
$M_\odot$	the mass of the Sun – $M_\odot \simeq 2 \times 10^{33}\,\mathrm{g}$, 48
AGASA	ultra-high energy cosmic-ray array in Japan, 163
AGN	active galactic nucleus – galaxy with an unusually bright nucleus whose emission is due to a massive central black hole, 68
ANITA	a balloon experiment which flew in Antarctica – sensitive to neutrinos of energy $\gtrsim 10^{18}\,\mathrm{eV}$ and above, 121
ANTARES	a Cherenkov neutrino VHE neutrino detector in the Mediterranean south of France – one of the forerunners of the planned cubic kilometer KM3NeT, 121
ASCA	Japanese X-ray satellite, 127
ATIC	a cosmic ray and dark matter space experiment, 182
ATLAS	A Large Toroidal Apparatus experiment on the LHC, 189
AUGER	the Pierre Auger Cosmic Ray Observatory in Argentina, 121
axion	a type of candidate dark matter particle, 186
AXP	anomalous X-ray pulsar, 96
BAT	burst alert telescope onboard the Swift satellite, 104
BATSE	Burst and Transient Spectroscopy Experiment onboard the Compton Gamma Ray Observatory (CGRO), 104
BBN	Big Bang Nucleosynthesis, 41
Beppo-SAX	Italian–Dutch X-ray/UV satellite which first discovered gamma-ray burst afterglows, 104
BH	black hole, 54

BL Lac | BL Lacertae – an AGN belonging to the sub-family of blazars, 80

BLRG | broad line radio galaxy, 74

BSM | beyond the Standard Model – referring to ideas and models outside of the current Standard Model of particle physics, 6

CANGAROO | "Collaboration between Australia and Nippon for a Gamma Ray Observatory in the Outback": array of air Cherenkov telescopes in Australia, 5

CAST | CERN Axion Solar Telescope, 186

CDMS | a dark matter direct search experiment, 184

CGRO | Compton Gamma Ray Observatory, 125

Chandra | Chandra X-ray Observatory – a NASA cornerstone mission observing in the X-ray range, 85

Cherenkov effect | used to measure energetic particles as well as neutrinos or photons through softer optical secondary photons emitted by them, 90

CMB | cosmic microwave background radiation, 39

CMS | Compact Muon Solenoid experiment on the LHC, 189

DAMA | a dark matter direct search experiment, 184

DUSEL | Deep Underground Science and Engineering Laboratory – approved for deployment in the Homestake mine in South Dakota, depths of between 4 and 6 km, 6

EIC | external inverse Compton process where the photons being upscattered come from outside the region containing the scattering electrons, 130

ESA | European Space Agency, 59

FERMI | Fermi Gamma Ray Space Observatory (formerly GLAST) – a satellite launched by NASA in 2008, 5

FermiLab | Fermi National Accelerator Laboratory – near Chicago (USA), 193

FR | Fanaroff–Riley – a term used to classify radio galaxies, 74

FSRQ | flat-spectrum radio quasar, 80

GBM   Gamma-ray Burst Monitor – an all-sky burst monitor on
                   the Fermi spacecraft sensitive in the 8 keV–30 MeV
                   energy range, 118
Geminga   name of a nearby pulsar, 182
Gran Sasso   Gran Sasso National Laboratory – underground
                   laboratory for particle astrophysics of the Italian
                   National Institute of Nuclear Physics, 6
GRB   gamma-ray burst source, 57
GUT   Grand Unified Theories – collective name for theories
                   beyond the Standard Model of particle physics which
                   attempt to consolidate the electroweak and the strong
                   interactions in a single unified theory, 6
GW   gravitational wave, 64, 143
GZK   Greisen–Zatsepin–Kuzmin – referring to photo-meson
                   interactions of cosmic-ray protons above 5×10^{19} eV
                   interacting with cosmic microwave background photons
                   which imposes a cutoff in the proton spectrum, 158

HESS   High Energy Stereoscopic System array of five air
                   Cherenkov telescopes in Namibia, 5
HETE-2   NASA satellite which played a significant role in
                   following up gamma-ray burst afterglows in the period
                   preceding Swift, 104
HMXB   high mass X-ray binary, 94
Homestake   mine in South Dakota housing the Davis neutrino
                   experiment, 167

IACT   imaging air Cherenkov telescope array consisting of
                   multiple telescopes functioning as an interferometer to
                   image the Cherenkov radiation produced by $\gtrsim 0.1$ TeV
                   gamma-rays from astrophysical sources, 5
IC   inverse Compton process whereby lower energy
                   photons are upscattered in energy by higher energy
                   electrons, 80
IceCube   cubic kilometer neutrino Cherenkov detector at the
                   South Pole, 5
IR   infrared, 134

KEK   Japan National Accelerator Laboratory, 192
KIMS   a dark matter direct search experiment, 184

KM3NeT	cubic kilometer neutrino Cherenkov detector being planned in the Mediterranean sea, 5
kpc	kiloparsec – a distance equivalent to 3×10^{21} cm, 70
LAT	Large Area Telescope onboard NASA's Fermi spacecraft – sensitive in the 20 MeV–300 GeV energy range, 118
LBL	low peak blazar – the second spectral peak extends up to X-ray or MeV energies only, 129
LCDM	Lambda Cold Dark Matter, 48
LHC	the Large Hadron Collider is the largest particle accelerator in the world – starting operations in 2009 at CERN in Switzerland, 5
LIGO	Laser Interferometric Gravitational Observatory – a gravitational wave detector array deployed in Washington and Louisiana states (USA), 8
LISA	Laser Interferometer Space Antenna – a gravitational wave detector spacecraft scheduled for launch in 2015 by the European Space Agency and NASA, 8
LMC	Large Magellanic Cloud – a nearby small satellite galaxy of our Milky Way, 97
LMXB	low mass X-ray binary, 94
MACHO	massive compact halo object, 45
MAGIC	Major Atmospheric Gamma-ray Imaging Cherenkov Telescope is a set of two very large (17 m diameter) Cherenkov telescopes in the Canary Islands, 5
MBH	massive black hole – referring to black holes with masses in excess of $\sim 10^4$–10^5 solar masses, 54
MBR	microwave background radiation (same as CMB), 134
MSW	Mikheyev–Smirnov–Wolfenstein matter neutrino oscillation effect, 169
MW	Milky Way – the name of our galaxy, 69
NASA	National Aeronautics and Space Agency (USA), 59
NEMO	a Cherenkov neutrino VHE neutrino detector in the Mediterranean near Sicily in Italy – another forerunner of the planned cubic kilometer KM3NeT, 121
NESTOR	a Cherenkov neutrino VHE neutrino detector in the Mediterranean near Pylos in Greece – another forerunner of the planned cubic kilometer KM3NeT, 121
NGC	New General Catalog – a standard catalog of galaxies, 71

NLRG	narrow line radio galaxy, 74
NS	neutron star, 65
PAMELA	a cosmic ray and dark matter space experiment, 182
pc	parsec – a distance equivalent to 3×10^{18} cm, 80
PSR	abbreviation of pulsar – used to label pulsar sources, 125
PWN	pulsar wind nebula, 126
QCD	quantum chromodynamics – the standard model theory of strong interactions, 23
QG	quantum gravity, 119
QPO	quasi-periodic objects – a type of astrophysical source whose X-ray emission exhibits a periodicity which is not well defined and typically varies, 99
QSO	quasi-stellar object – the name of the most energetic type of active galactic nuclei, 79
RHIC	Relativistic Heavy Ion Collider – at Brookhaven National Laboratory in the USA, 194
SD	surface detector (of cosmic rays), 163
SGR	soft gamma repeater – a very high field magnetized neutron star which has repeated gamma-ray flare-up episodes, 97
SN	supernova – including type I (SN I) and type II (SNII), 86
SNO	Sudbury Neutrino Observatory in Canada, 167
SNR	supernova remnant, 83
Soudan	Soudan Underground Laboratory in Minnesota, 192
SSC	synchrotron self-Compton process where the same electrons producing synchrotron photons upscatter these to higher energies via inverse Compton scattering, 130
Super-Kamiokande	(super-K) – an underground neutrino detector in the Japanese Kamioka mine, 6
SUSY	super-symmetric grand unified theories – also SUSY GUTs, 28

Suzaku	Japanese hard X-ray satellite sensitive from 0.2 to 600 keV, 85
Swift	a NASA satellite covering the gamma-ray through X-ray to UV/optical energies – designed to investigate gamma-ray bursts and to follow up promptly their afterglows, 88
TNT	trinitrotoluene – a high explosive, 56
Tokai	Japanese university housing a large accelerator, 192
UHE	ultra-high energy – generally referring to energies $\gg$ TeV, 83
UV	ultraviolet, 85
UVOT	UV/optical telescope onboard the Swift satellite, 104
VERITAS	Very Energetic Radiation Imaging Telescope Array System – set of four air Cherenkov telescopes in Arizona, 5
VHE	very high energy – referring to energies in the GeV to TeV range, 80
VIRGO	Virgo Gravitational Wave Detector at the European Gravitational Observatory near Pisa in Italy, 8
VLA	Very Large Array – a large radio interferometer array in New Mexico, 77
VLBA	Very Large Baseline Array – a radio interferometer array encompassing various sites across the USA, 71
WD	white dwarf star, 85
WIMP	weakly interacting massive particle – the generic name for a class of particle which is thought to provide the dark matter in the Universe, 5
WMAP	Wilkinson Microwave Anisotropy Experiment, 187
XENON10	a dark matter direct search experiment, 184
XMM	X-ray Multi-Mission experiment – an ESA large X-ray space observatory, 85
XRT	X-ray telescope onboard the Swift satellite, 104
ZEPLIN-II	a dark matter direct search experiment, 184

Table A.1. *Relations between commonly used astrophysical c.g.s., and SI units*

Type	Symbol (name)	c.g.s.	SI
weight	g (gram)	g	10^{-3} kg
time	s (second)	s	s
time	yr (year)	3×10^7 s	3×10^7 s
length	cm (centimeter)	cm	10^{-2} m
energy	erg	erg $= \text{g cm}^2\,\text{s}^{-2}$	10^{-7} joule
energy	eV (electronvolt)	1.602×10^{-12} erg	1.602×10^{-19} joule
energy	keV (kilo-eV)	1.602×10^{-9} erg	1.602×10^{-16} joule
energy	MeV (mega-eV)	1.602×10^{-6} erg	1.602×10^{-13} joule
energy	GeV (giga-eV)	1.602×10^{-3} erg	1.602×10^{-10} joule
energy	PeV (peta-eV)	1.602 erg	1.602×10^{-7} joule
energy	EeV (exa-eV)	1.602×10^3 erg	1.602×10^{-4} joule
force	dyne	g cm s^{-2}	10^{-5} newtons
power (luminosity)	erg s^{-1}	erg s^{-1}	10^{-7} watt
speed of light	c	$2.998 \times 10^{10}\,\text{cm s}^{-1}$	$2.998 \times 10^8\,\text{m s}^{-1}$
solar mass	$M_\odot$	1.989×10^{33} g	1.989×10^{30} kg
solar luminosity	$L_\odot$	$3.826 \times 10^{33}\,\text{erg s}^{-1}$	3.826×10^{26} watt
Sun–Earth mean distance	A.U. ("astronomical unit")	1.496×10^{13} cm	1.4966×10^{11} m
distance (astron.)	pc (parsec)	3.086×10^{18} cm	3.086×10^{15} m
distance (astron.)	kpc (kiloparsec)	3.086×10^{21} cm	3.086×10^{18} m
distance (astron.)	Mpc (megaparsec)	3.086×10^{24} cm	3.086×10^{21} m
Hubble constant	H_0	$\sim 74 \pm 4$ (km/s)/Mpc	$\sim 74 \pm 4$ (km/s)/Mpc
Hubble radius	c/H_0	$\sim 4\,\text{Gpc} = 1.2 \times 10^{28}$ cm	$\sim 1.2 \times 10^{22}$ km
Hubble time	$1/H_0$	$\sim 4 \times 10^{17}$ s	$\sim 1.33 \times 10^{10}$ yr

Index